Digital Video
in the PC Environment

Digital Video
in the PC Environment
Second Edition

Arch C. Luther

Intertext Publications
McGraw-Hill Book Company

New York St. Louis San Francisco Auckland Bogotá
Hamburg London Madrid Mexico Milan Montreal
New Delhi Panama Paris São Paolo
Singapore Sydney Tokyo Toronto

Library of Congress Catalog Card Number 88-83067

10 9 8 7 6 5 4 3 2 1

ISBN 0-07-039176-9 [Hardcover]
ISBN 0-07-039177-7 [Softcover]

DVI screen for cover created by Exhibit Technology, Inc., New York, NY Image courtesy of NASA.

Macintosh is a registered trademark of Apple Computer, Inc.
PC/AT, Infowindows, PS/2 are registered trademarks of IBM Corporation.
GEM is a trademark of Digital Research, Inc.
Transputer is a trademark of Inmos.
DVI, ActionMedia, 80286, 80386 are trademarks of Intel corporation.
MS-DOS and MS-Windows are trademarks of Microsoft Corporation.
Betacam and VIEW are trademarks of Sony Corporation.
Targa is a registered trademark of Truevision, Inc.
Authology: Multimedia is a registered trademark of CEIT Systems, Inc.
Lumena is a trademark of Time Arts, Inc.
MEDIAscript is a trademark of Network Technology Corporation.

Intertext Publications/Multiscience Press, Inc.
One Lincoln Plaza
New York, NY 10023

McGraw-Hill Book Company
1221 Avenue of the Americas
New York, NY 10020

In memory of
Arch "Clint" Luther, III
1970-1989

Contents

Preface

In early 1984 when I was making the move from RCA Broadcast Systems Division at Camden, New Jersey, to the RCA Laboratories at Princeton, New Jersey, I had no idea where it was going to take me, except I was sure it would be *different*. I would never have guessed that I was becoming involved in what would turn out to be the most exciting project of a long career that included many other exciting projects, or that it would lead to early retirement from RCA, and now the freedom to pursue consulting, software design, and writing. But those are the steps which bring me to writing on this page.

In that move from Camden to Princeton I left the television broadcast equipment business and joined a project which had the objective of putting television-style video and audio on a personal computer. The result of that project, DVI Technology, has captured the imagination of both the personal computer and the interactive video industries. Bringing together audio, video, and personal computers creates a system, which can fulfill a large number of needs for information systems with vast storage capacity, and that is user-friendly and can deliver any type of presentation to the user.

However, the development of an exciting technology does not alone guarantee its application — it takes a community of people and companies who know what users want, and who will learn enough about the new technology to see how applying that technology will provide better answers to real problems. Because of the broad collection of techniques and skills needed to work with a system like DVI Technology, people who want to develop systems or software will need to learn a combination of techniques, each of which now exists in diverse industries. There are few people who already have the correct collection of skills. This book is addressed to those who are interested in learning about this exciting new technology and the special skills it takes to apply it.

DVI Technology makes possible the first systems which truly merge personal computers and television. The book begins by introducing the need for such systems, and then discusses the several technologies which must go into such systems. Then DVI Technology is introduced and becomes an illustrative system for all of the discussion that follows. This will give you a good understanding of the fundamentals for such systems so you should be able to evaluate all competitive offerings. Because DVI Technology is covered in detail, you will get a complete picture of its capabilities and how it can be applied.

The list of skills needed to work with an audio/video/computer system is very long. Here are some of the disciplines required:

Audio engineering
Video engineering
Computer engineering
Audio/video production
Audio/video postproduction
Artistic design
Graphic art
Script writing
Creative writing
Publishing
Programming
Image processing

The above skill areas are all well developed for other fields, such as television, motion pictures, theater, audio records, paper publishing, or personal computers. In this book, I will tap into that knowledge and try to extract the parts which will be needed for audio/video/computer systems, and present them all in one place. One book cannot possibly give you comprehensive knowledge in all these areas, but what I hope it will do is show you how they all come together in the creation of systems and applications. I also have tried to include references (which are at the end of the book) to take you further into all of the areas. The book is intended to teach *what* things do, rather than *how* they do them, and *what* you need to do to apply the features of audio/video/computer systems to your applications.

In writing a book, the author has to know who the audience is, and what their background is like. It is not possible to start everything out from zero — there must be some assumptions about what the reader knows when he or she opens the book for the first time. With regard to background, I will assume that the reader has a technical background and interest, is familiar with electronic systems and personal computers including some kind of programming, but is not necessarily skilled in any of the disciplines listed above.

A word is necessary about the need for programming knowledge in reading this book. The most valuable feature about any computer-based digital system is that it should be *programmable*. Therefore, it is the programmers who will make that system perform, and they are the people who need to know and understand the system most thoroughly. The book has been written to fulfill that need; but that does not mean that a non-programmer should get lost in the material. Many of the people who will be involved in the process of developing audio/video/computer applications will not be programmers (just look at the list above), and therefore I have carefully segregated the deep software discussions into separate sections which a non-programmer may safely skip over. At the end of each of these sections I have included a "points made by the software example" paragraph which will bring the non-programmer back up to speed to continue the book.

The diverse industries involved here have a lot of their own jargon. One step necessary for becoming involved in this subject is to learn enough of that jargon so you can understand the specialists and talk to them. That is a major objective of the book, and I have tried rigorously to define all the jargon as we go along. New words are italicized the first time they are introduced, and they are all included in a Glossary at the end—Appendix C. However, once I have introduced and defined a word, I will then use it wherever needed and I will assume that you now understand it! So watch carefully for italicized words and store away their definitions — chances are you'll run into them later.

In order to be as rigorous as I just promised, we should stop here and define the jargon used in the title of this book — *Digital Video* and *PC*. For *digital video*, we have to start with *video* — in this book *video* refers to any system for electronic representation of images, whether real or computer generated. (*Real images* are images captured originally from nature, usually by photography, cinematography, or videography.) Therefore, video includes reproduction of not only the real images of television, but also the familiar computer display images, whether text, graphics, or pictures. *Digital* video refers to a system where all of the information that represents images is in some kind of computer data form, which can be displayed or manipulated by a computer. Digital video specifically excludes *analog* video, where images are represented by continuous-scale electrical signals.

Everyone probably knows that *PC* means *personal computer* — but what *is* a personal computer? In this book, a personal computer is a software-controlled system which contains general-purpose computer functions, is intended to be used by one person, and is low enough in cost to be widely deployed (dollar cost of a base unit in the low thousands). There also are higher-priced (five figures and up) *workstation* products intended mostly for engineering or scientific work — these are not viewed as personal computers in this book.

The PC *environment* refers to everything surrounding a personal computer — the PC itself, its peripherals, and its software. All of these will be affected by the introduction of any exciting new capability such as DVI Technology, and therefore we must treat the environment as a whole.

Being involved in the development of DVI Technology, and then having the opportunity to be the first one to tell its story in a full-length book has been a great thrill for me. I hope you will feel that same thrill as you learn about DVI Technology and begin to realize the full possibilities for application of this outstanding new technology.

Arch C. Luther
Merchantville, New Jersey
September 1988

Preface to the Second Edition

Nearly two years have passed since I wrote the First Edition of this book. Those have been two very full years for the multimedia industry, for DVI Technology, and for me. Two years ago, the word *multimedia* was just coming into use to refer to what I called "audio-video-computer systems" in the First Edition. Today the word multimedia is the name of a budding industry and it appears everywhere — in publications, in product names, and in conversations around the world.

DVI Technology has also come a long way in two years. Two years ago, Intel Corporation had just acquired the technology, and had just introduced the first products. Today there are second-generation products of hardware and software, and we know that the third-generation is coming along in the design center and in the laboratory. IBM has become a partner in the technology, and DVI products for the PS/2 product line are available. Some vertical-market software developers have been working with DVI Technology for more than two years now, and their efforts are beginning to show fruit as finished applications in their fields of specialty. Other application developers have adopted the technology and are broadening its thrust in many fields.

Application development for DVI Technology is becoming easier because of the introduction of authoring software for the technology. Two authoring systems are on the market, and others are under development by several companies. The availability of authoring will increase the rate of DVI Technology application development.

Standards for several aspects of multimedia are under consideration by industry groups. It's too early to report any conclusions yet, but the proponents of DVI Technology are providing a strong presence. The software-based nature of the technology insures that through revision of software it will be able to respond to whatever standards decisions are eventually reached.

As for myself, I have continued to devote my full time (and more) to activities related to DVI Technology. As a consultant for Intel and IBM, I have had the opportunity to participate in much of the continued industry excitement for the technology. I have also been able to market the authoring software that I developed for my own use over the last three years, by an agreement with Network Technology Corporation of Springfield, VA. This has led to MEDIAscript™, one of the two currently available DVI authoring systems.

Multimedia is a dynamic, exciting, and growing field. I believe that the second edition of this book will convince you of that, and the possibilities for this exciting technology have only become greater in the last two years.

Arch C. Luther
Merchantville, New Jersey
June, 1990

Acknowledgments

This book would not have been possible without the experience I gained by being a part of the development team for DVI Technology. That team was headed by Arthur Kaiman, Director of Digital Products Research at the David Sarnoff Research Center. Art, more than anyone else, was responsible for finding ways to keep the project going through many rough spots in the corporate environment of RCA, then GE, and then the David Sarnoff Research Center, a subsidiary of SRI International. The contribution of Richard Stauffer also must be acknowledged — Rick came into the DVI project from GE after the GE acquisition of RCA. He saw the potential of the technology and became the principal spokesperson to carry the DVI torch in GE. Now, the torch has been passed to the Intel Corporation, where Dave House, President, Microcomputer Components Group and the more than 100 people (including those mentioned above) of the Intel Princeton Operation at Plainsboro, New Jersey have accepted the challenge — and the opportunity — to make DVI Technology the multimedia standard for desktop computers. And lastly — but actually first — I want to recognize Larry Ryan: he is the person who first conceived the technology and who sold all the rest of us on the idea. Then Larry kept us on the track and dogged us all until we had made it work. He is truly the father of DVI Technology.

The DVI development group.

The photograph on the preceding page, taken September 6, 1988 shows the group of people who developed DVI Technology at the David Sarnoff Research Center, most of whom are now bringing the technology to market as employess of Intel Corporation. From left to right, beginning in the back row, they are: Lou Lippincott, Bill Gallas, Dave Ripley, Al Simon, Tom Craver, Holly Faubel, Dave Sprague, Rick Stauffer, Mike Buondonno, Stu Golin, Doug Dixon, Bob Winder, Al Procassini, Dave Lippincott, Harold Hanson, Larry Ryan, Bill Clem, Kevin Harney, Don Sauer, Joe Lala, John Egan, Sanjay Vinekar, David Nimrod, Mike Patti, Wayne Smith, Marc Yellin, Sandra Morris; (front row) Arch Luther, Mike Keith, Gary Lavelle, Fred Vannozzi, Sal Noto, Paula Zimmerman, George Breen, Jim Jeffers, Skip Kennedy, Art Kaiman, Irene Hashfield, Cindy Young, Carolyn Perlman, Al Korenjak, Mike Tinker.

Any project such as this book requires the help of many others to assist the author in accomplishing his objective. The following people gave freely of their time in offering information and suggestions, and in reviewing the many versions of the manuscript which it took to get to the finished First Edition: Mark Bunzel, Tom Craver, Douglas Dixon, Irene Hashfield, Bob Hurst, Art Kaiman, Clint Luther, Larry Ryan, Alan Rose, Al Simon, Fred Vannozzi, and Paula Zimmerman.

I also want to give special recognition to Kayle Luther (my daughter) who joined me recently as an associate and contributed extensively to the completion of the Second Edition.

1

Introduction

Television is an integral part of daily life in most of the world today. There are thousands of broadcasting stations and hundreds of millions of receivers in use. In addition, more than two hundred million VCRs bring television to viewers at times and places and with subject material that is not reached by broadcasting. We all accept television as a medium that can bring us realistic pictures and sound, capable of absorbing our attention and our emotions. Television has clearly changed the world we live in.

The personal computer is also an integral part of life for many of us — in the office, in the school, and in the home. We use PCs as a tool in business, to train and teach, and to entertain us. With a PC we are able to perform useful tasks such as numerical calculations, capturing and formatting of text, and drawing for design or artistic purposes. The PC displays pictures and makes sounds which are appropriate for the PC's purpose even though we usually would not say that they are realistic. Under our control from keyboard, mouse, or tablet, the computer is also clearly changing our world.

But now there is a product that brings television and personal computers together — a new digital product which combines the controllability of personal computers with the realism of television images and sounds. This marriage is possible because of the advances of digital video and audio technologies and the development of optical digital storage systems. This book is about such products, which combine personal computers with digital video and audio — *multimedia* computers.

Two Worlds — Analog and Digital

These two product lines — television and personal computers — have common origins in electronic technology, but they have developed independently, almost as though they were in different worlds. Television's growth has centered around broadcasting and its auxiliary uses, and the personal computer is an extension of larger computers, which were originally developed for mathematical analysis and office financial uses. It is not the purpose here to explain why it happened that way, but simply to recognize that two separate and independent industries have developed. This separation, caused by fundamentally different market objectives, has often led to the same function being developed in the two industries using different techniques — this is the case for pictures and sounds. With respect to pictures and sounds, television is largely an analog technology, whereas PCs are totally digital.

Television was designed to reproduce natural images and sounds, and nature is basically analog. In nature, the brightness of an object can have any value, and the object can be shaded with minute gradation. Likewise, a natural object can take on any position that is consistent with the action of gravity, and it may be moved smoothly with microscopic precision. Real images and sounds thus are made up of intensity values and positions that can have any level — continuous functions in both space and time — that is, analog. However, there is another reason that television is analog: at the time television was being developed, it was exceedingly difficult (and expensive) to reproduce images and sounds with any kind of digital technique. Today, this situation is quite different — we have low-cost digital integrated circuits costing a fraction of a dollar which can do digital functions that cost thousands of dollars to do with vacuum tubes in the 1950s. This is a major reason why digital video is now a possibility for low-cost systems.

Personal computers were originally designed to work with numbers;

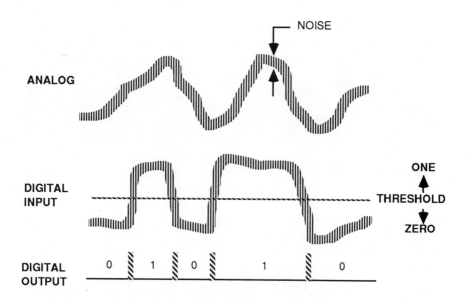

Figure 1.1   Analog and digital signals with noise

they are digital because a digital system can represent numbers with any desired degree of accuracy, not limited by any characteristics of the hardware. By being digital, the personal computer is able to get around an inherent limitation of all electronic circuits — electronic signal reproduction is never exact. In any electronic system, there is always a noise level and a distortion level that will cause minute changes to any signals passing through or being processed by the electronics. If that signal is an analog representation of something like light intensity or sound level, small errors will be introduced into the signal. These are most commonly observable as hiss in sound and as snow in pictures. For picture and sound reproduction, these errors can be made small enough so that they are not disturbing. However, those same errors would be unacceptable to any extent in a computer used to process numbers that, for example, represent a multimillion dollar payroll. The basic concept of a computer is that the hardware should not make errors. (Computer errors can occur because of improperly designed software, but once these are found and corrected, they should never recur.) The solution to making perfect computer hardware is to do it digitally. Digital technology uses signals that have only a few significant values (most systems

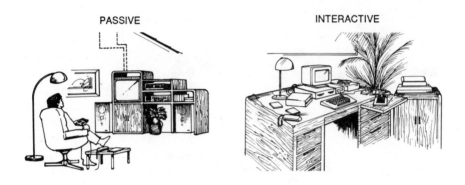

Figure 1.2   Passive viewing – the living room scenario. Interactive
                    viewing – the desktop scenario.

are *binary*—two levels). By recognizing only two levels, a binary digital system makes insignificant the inherent noise and distortions of electronic circuits, and the two levels can be reproduced exactly. Figure 1.1 shows analog and digital signals with a small noise level superimposed. The digital example shows that by using a *threshold* in the digital circuit, where any level above the threshold is called 1 and any level below the threshold is called 0, two levels are handled without the circuit noise having any effect.

Since traditional use of computers for number-intensive tasks requires freedom from errors, the computer industry has developed the idea that errors should never be tolerated in computers, and much technology has been developed to achieve this — including ways to produce pictures and sounds with no errors. However, producing truly realistic pictures and sounds with pure computer techniques still proves to be expensive, which is why most computers do not have realistic pictures and sounds. At the same time, the television industry has learned to live with hardware that always makes errors, and many concepts for controlling or "hiding" the errors have been developed. Bringing real video and audio into the computer world becomes possible at low cost by use of some of television's "hiding" tricks in computer images and sounds. This allows the development of digital algorithms, which let a computer process and display realistic motion video and audio just as effectively as it can deal with numbers.

Market Trends Bring TV and PCs Together

But let's return to the market trends that are leading television and computers to the altar. Why would anyone want a product that contains both of these very different technologies, anyway? The answer is found largely in one word — *interactivity* — which is a major characteristic of computers, but not of television. Interactivity means control by the user, but in a much more significant sense than our "control" of a television set, which consists primarily of selection from a limited number of predetermined choices (channels). With television, the user selects the material to be shown, but then he or she simply sits back and watches passively as that material is presented (Figure 1.2). This is a satisfactory mode for entertainment, but it becomes seriously limiting if we are trying to use the television for other purposes such as teaching, training, or selling. There is no way for the user to affect the flow of a television program once it has been chosen other than by selecting another program.

On the other hand, a personal computer is almost always interactive. Once a computer program has been selected and started, the computer expects the user to interact with it, providing input and making decisions. This is true whether the program is a word processor where the input is from the keyboard, or a paint program, which might be controlled from a mouse, or a game program controlled by a joystick. Because of the inherent interactivity of personal computers, a whole stable of input/output (I/O) devices has been developed for control and for output from a computer. In such a rich interactive environment, video and audio are simply two more kinds of output that we can get from a computer — and we will expect to be able to control them just like any other computer output.

Who Needs Interactive Audio/Video?

So, we can make video and audio interactive by putting them on a personal computer — who needs that? People who are teaching, training, or selling—that's who. This also includes people who are doing other applications, but for whom teaching, training, or selling is part of their application or could be part of the application if it were easy to do. A good example of the latter is a computer program for an auto parts catalog. How much more effective such a program would be if it also included animated exploded views of the parts, and even video and audio demonstrations of parts installation procedures. The following paragraphs explain a variety of applications for multimedia systems.

Surrogate travel

A large class of interactive audio/video applications are covered by the technique called *surrogate travel*. With surrogate travel, we can sit in our living room or our office and ask the computer to take us to a distant site and show us the scenes and let us hear the sounds of that locality while we interactively control our position within the site. This is particularly effective for a situation where it is physically or economically impossible for the actual travel to occur. For example, one application developed during the research leading to DVI Technology takes the user to the Mayan ruin site at Palenque, Mexico. The application — called *Palenque* — was a joint project with the Bank Street College of Education in New York City, and it was a spinoff from Bank Street's PBS educational TV program *Voyage of the Mimi. Palenque's* surrogate travel uses a photographic technique where a photographer carries a movie camera rig that takes one frame each time the photographer takes a step — this captures a frame about every three feet. Frames were taken that covered several miles of paths among the ruins, inside the buildings, and in the nearby rain forest. The user may call up those frames and essentially walk around the site. We can go forward, backward, or take paths left or right, and all the time the computer shows what we would see if we were actually at Palenque. The computer also presents us with the sounds of the site, using a library of audio captured on the site and presented to us as we move about. The effectiveness of this technique has to be experienced to be appreciated.

Another field requiring the distribution of all kinds of information is the real estate field. If you were planning to move to a distant city, wouldn't it be valuable to be able to use surrogate travel to explore all the suburbs and living areas of that city while seated at your computer — before you went to the city — in order to narrow down the areas where you might like to live? Interactive audio/video can do that, and it can go still further and show you information about all of the houses for sale in the areas you are interested in. So, you can do a lot of your planning for a house-hunting trip without ever getting on a plane!

Synthetic video

A computer audio/video system also is capable of creating *synthetic video* images, constructed by combining a computer three-dimensional model with real video images of surfaces, patterns, and textures. This technique allows a range of applications that create realistic rendering of computer-designed objects, interior designs, landscapes, or architecture. For example, an interior design application, which also began during the research for DVI Technology, allows an interior designer to create a model of a room and display realistic views of that room. This application — called *Design & Decorate* — is being continued by Design & Decorate, Inc., a company in New York City which has been formed to pursue this technology. Working at the video screen, the designer selects furniture pieces from a large electronic catalog, positions them in the room, chooses fabric covering from a fabric catalog, adds wallcovering selections, carpets and paints, and other decorative objects. Then in a matter of seconds, the computer constructs a realistic perspective rendering of that room from any point of view. This can be used to show the designer's client what the room design will look like with all the selections the designer and client have made.

The synthetic video technique can also be used to train operators of vehicles such as military tanks, commercial trucks and off-road vehicles, or airplanes. In this case, the 3-D rendering must be optimized for speed, so that the rendered view can be presented several times per second as the operator moves his vehicle around in a computer-simulated environment.

Other applications

The medical field has many applications for interactive audio/video. The problem of keeping doctors up to date with the progress of medicine without taking too much of their valuable time has prompted many people to pursue the use of electronic media to distribute information about new developments. This requires audio, video, text, and numerical data accessible by the doctor, depending on the subject of his or her interest at a particular time. An interactive audio/video system can fill this need.

Interactive video and audio have also been proposed for use in teaching theatrical production and design, for museum displays of various subjects, for management presentations, and many other purposes. It is clear that as this kind of technology becomes widely available and lower in price, the applications will extend far beyond what we can see now.

Doing Interactive Video with a Video Disc

The previous examples show the range of applications for a computer audio/video system. In fact, many of those applications are available; systems have already been assembled by combining several existing technologies. For the most part, this is being done using the laser video disc as the medium for the video and audio; the laser video disc is interfaced to a computer for control purposes. However, because the laser disc is analog for video and audio, the video and audio outputs go directly to a monitor and speakers, and the output cannot be controlled or manipulated by the computer. Therefore, user control is largely limited to a selection process, and the only video and audio that can be presented is whatever the designer of the video disc chose to put onto the disc. In spite of that limitation, for applications that do not require too much video and audio, the disc designer can include all of the necessary combinations on the disc and produce an effective application. There is a growing marketplace for video disc plus computer systems which has led to several manufacturers producing packaged systems. Examples are the Sony View system, the IBM Infowindows system, and the EIDS system built for the U.S. Army.

What's So Great about All-Digital?

It would be much better if the system were *all-digital*. By all-digital I mean that everything — audio, video, data — is stored digitally and is processed through the computer for presentation. Then the same digital medium can be used for storage of all classes of information needed for an application, and only one type of storage drive will be needed in a system. Furthermore, the processing power and programmability of the computer can be used on the audio and video to increase the variety of ways in which audio and video can be stored and displayed. An even further capability uses the computer to create its own video or audio — producing synthetic video and audio, presented instead of or along with the real video and audio from the storage medium.

As we will see in later chapters, creating an all-digital video/audio/ computer system is not an easy task, particularly at a hardware cost low enough to create mass markets for such products. However, the people at the David Sarnoff Research Center in Princeton, New Jersey, took on that challenge in the early 1980s, and they developed Digital Video Interactive (DVI) Technology, which does it all! DVI Technology is owned by Intel Corporation, who has continued to develop the technology and is offerring it to a wide range of markets through their Intel Princeton Operation. DVI technology is presented in this book as the example of an all-digital audio, video, and computer system. Other systems that are being developed for related objectives will be mentioned as we go along.

2

Analog Video Fundamentals

In order to pursue the discussion of digital video, we need to understand the principles of analog video as that exists in the television industry. In most cases we will be using standard television equipment and systems to originally produce our video — converting it to digital video will be a later step. Therefore a good knowledge of analog video nomenclature, characteristics, performance, and limitations will be essential to our appreciation of the overall video system. This chapter will explain analog video fundamentals as they relate to the uses of video in digital formats. It presents a simplified model of an analog video system, which will probably seem elementary to readers who are already familiar with video. However, this chapter is intended for a reader who does not have a video background, and it contains only enough to help that reader get through the rest of this book. You will be introduced to a lot of video and television terminology that will be used extensively later in this book. So, for those who need to know something about video — let's go!

We have already established that most things in nature are analog. Real images and sounds are based on light intensity and sound pressure

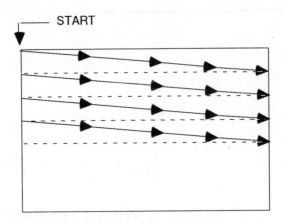

Figure 2.1   **Raster scanning of an image**

values, which are continuous functions in space and time. For television we must convert images and sounds to electrical signals. That is done by appropriate use of *sensors*, also called *transducers*. Sensors for converting images and sounds to electronic signals are typically analog devices, with analog outputs. The world of television and sound recording is based on these devices. Video cameras and microphones (the sensors) are familiar objects to almost everyone, and their purpose is generally well understood. Here, however, we will concentrate on how they work. This chapter is concerned with video; Chapter 4 will take up audio.

Raster Scanning Principles

The purpose of a video camera is to convert an image in front of the camera into an electrical signal. An electrical signal has only one value at any instant in time — it is one-dimensional, but an image is two-dimensional, having many values at all the different positions in the image. Conversion of the two-dimensional image into a one-dimensional electrical signal is accomplished by *scanning* that image in an orderly pattern called a *raster*. With scanning, we move a sensing point rapidly over the image — fast enough to capture the complete image before it moves too much. As the sensing point moves, the electrical output changes in response to the brightness or color of the image under the sensing point. The varying electrical signal from the sensor then represents the image as a series of values spread out in time — this is called a *video signal*.

Figure 2.1 shows a raster scanning pattern. Scanning of the image begins at the upper left corner and progresses horizontally across the image, making a *scanning line*. At the same time, the scanning point is being moved down at a very much slower rate. When the right side of the image is reached, the scanning point snaps back to the left side of the image. Because of the slow vertical motion of the scanning point, it is now a little below the starting point of the first line. It then scans across again on the next line, snaps back, and continues until all of the image has been scanned vertically by a series of lines. During each line scanned, the electrical output from the scanning sensor represents the light intensity of the image at each position of the scanning point. During the snap-back time (known as the *horizontal blanking interval*) it is customary to turn off the sensor so a zero-output (or *blanking level*) signal is sent out. The signal from a complete scan of the image is a sequence of line signals, separated by horizontal blanking intervals. This set of scanning lines is called a *frame*.

Aspect ratio

An important parameter of scanning is the *aspect ratio*. This is the ratio of the length of a scanning line horizontally on the image, to the distance covered vertically on the image by all the scanning lines. Aspect ratio can also be thought of as the width-to-height ratio of a frame. In present-day television, aspect ratio is standardized at 4:3. Other imaging systems, such as movies, use different aspect ratios ranging as high as 2:1 for some systems.

Sync

If the electrical signal from this scanning process is used to modulate the brightness of the beam in a cathode-ray tube that is being scanned exactly the same way as the sensor, the original image will be reproduced. This is what happens in a television set or a video monitor. However, the electrical signal(s) sent to the monitor must contain some additional information to ensure that the monitor scanning will be in synchronism with the sensor's scanning. This information is called *sync information*, and it may be included within the video signal itself during the blanking intervals, or it may be sent on a separate cable (or cables)

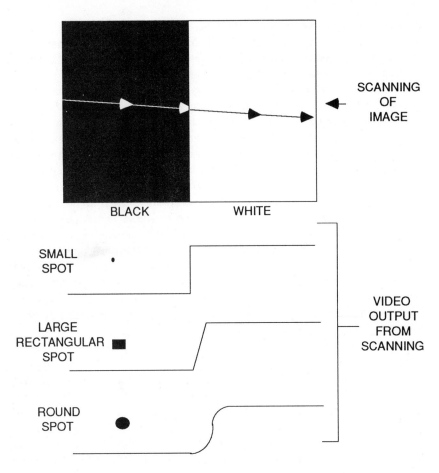

SCANNING
OF
IMAGE

BLACK   WHITE

SMALL
SPOT

LARGE
RECTANGULAR
SPOT

ROUND
SPOT

VIDEO
OUTPUT
FROM
SCANNING

**Figure 2.2   Scanning across a vertical edge with different sizes and
shapes of scanning spot**

just for the sync information.

Horizontal resolution

As the scanning point moves across one line, the electrical signal output
from the sensor changes continuously in response to the light level of the
part of the image that the sensor sees.   One measure of scanning
performance is the *horizontal resolution* of the pickup system, which
depends on the *size* of the scanning sensitive point. A smaller sensitive

point will give higher resolution. Figure 2.2 shows the result of scanning across a sharp vertical edge in an image using scanning sensors of different sizes. Note that the electrical output is zero while the scanning sensor is looking at the black area, and the output begins to rise as the sensor moves partially onto the white area. Full output (100) is reached when the sensor is completely on the white area.

To test the horizontal resolution performance of a system, which also measures the capability to reproduce horizontal fine detail, we place closely spaced vertical lines in front of the camera. If the sensor area is smaller than the space between the vertical lines, the lines will be reproduced, but when the sensor is too large the lines will average out under the sensor and will not be seen in the output signal. In the television business, horizontal resolution is measured by counting the number of black and white vertical lines that can be reproduced in a distance corresponding to the raster height. (The raster height was chosen as the standard basis for specifying television resolutions, both horizontal and vertical.) Thus a system that is said to have a horizontal resolution of 400 lines can reproduce 200 white and 200 black lines alternating across a horizontal distance corresponding to the height of the image.

Scanning across a pattern of vertical black and white lines produces a high-frequency electrical signal. It is important that the circuits used for processing or transmitting these signals have adequate *bandwidth* for the signal. Without going into the details of deriving the numbers, broadcast television systems require a bandwidth of 1 megahertz (MHz) for each 80 lines of horizontal resolution. The North American broadcast television system is designed for a bandwidth of 4.5 MHz, and this has a theoretical horizontal resolution limit of 360 lines. Bandwidth considerations affect the choice of scanning parameters, as we will see later.

Vertical resolution

The *vertical resolution* of a video system depends on the number of scanning lines used in one frame. The more lines there are, the higher is the vertical resolution. Broadcast television systems use either 525 lines (North America and Japan) or 625 lines (Europe, etc.) per frame. A small number of lines out of each frame (typically 40) are devoted to the *vertical blanking interval*. Both blanking intervals (horizontal and vertical) were originally intended to give time for the scanning beam in cameras or monitors to retrace before starting the next line or the next frame. However, in modern systems they have many other uses, since these intervals represent nonactive picture time where different information can be transmitted along with the video signal.

Frame rates for motion

For motion video, many frames must be scanned each second to produce an effect of smooth motion. In standard broadcast video systems, normal frame rates are 25 or 30 frames per second, depending on the country you are in. However, these frame rates — although they are high enough to deliver smooth motion — are not high enough to prevent a video display from having *flicker*. In order for the human eye not to perceive flicker in a bright image, the refresh rate of the image must be higher than 50 per second. However, to speed up the frame rate to that range while preserving horizontal resolution would require speeding up of all the scanning, both horizontal and vertical, therefore increasing the system bandwidth. To avoid this difficulty, all television systems use *interlace*.

Interlace in a television system means that more than one vertical scan is used to reproduce a complete frame. Broadcast television uses 2:1 interlace — 2 vertical scans for a complete frame. Larger interlace numbers have also been used in some special-purpose television systems. With 2:1 interlace, one vertical scan displays all the odd lines of a frame, and then a second vertical scan puts in all the even lines. At 30 frames per second (North America and Japan), the vertical rate is 60 scans per second. Since the eye does not readily see flickering objects that are small, the 30 per second repetition rate of any one line is not seen as flicker, but rather the entire picture appears to be being refreshed at 60 per second. (The use of computer-generated images, particularly graphics, sometimes causes difficulties in interlaced systems. This will be explained later.)

Sensors for TV Cameras

It is possible to make a television camera as described above with a single light-sensitive element; however that proves not to be an effective approach because the sensor only receives light from a point in the image for the small fraction of time that the sensor is looking at that point. Light coming from a point while the sensor is not looking at that point is wasted, and this is most of the time. The result is that this type of video sensor has extremely poor sensitivity — it takes a large amount of light to make a picture. All present-day video pickup devices use an *integrat-*

ing approach to collect all the light from every point of an image all the time. The use of integration in a pickup device increases the sensitivity thousands of times compared to non-integration pickup. With an integrating pickup device, the image is optically focused on a two-dimensional surface of photosensitive material that is able to collect all the light from all points of the image all the time, continuously building up an electrical charge at each point of the image on the surface. This charge is then read out and converted to a voltage by a separate process which scans the photosensitive surface.

Without going into all possible kinds of pickup devices, there are two major types in use today which differ primarily in the way they scan out the integrated and stored charge image. Vacuum-tube pickup devices (vidicon, saticon, etc.) collect the stored charge on a special surface deposited at the end of a glass vacuum tube. An electron beam scans out the signal in this kind of device. The other type of pickup device is solid-state, where the stored charge image is developed on a silicon chip, which is scanned out by a solid-state array overlaid on the same chip. The solid-state devices are known by names like CCD, MOS, etc. Both kinds of devices can have excellent performance, but there are many detail differences including cost, size, and types of supporting electronics required to interface them into a camera system.

Color Fundamentals

The previous discussion assumed that the image being scanned was monochrome. However, most real images are in color, and what we really want is to reproduce the image in color. Color video makes use of the *tri-stimulus theory* of color reproduction, which says that any color can be reproduced by appropriate mixing of three *primary colors*. In grade school we learned to paint colors by doing just that — mixing the three colors: "red," "blue," and yellow. (The use of quotes on "red" and "blue," but not yellow, is deliberate and will be explained below.) These paint colors are used to create all possible colors by mixing them and painting on white paper. This process is known technically as *subtractive* color mixing — because we are starting with the white paper which reflects all colors equally, and we are adding paints whose pigments filter the reflected white light to subtract certain colors. For example, we mix all three paint primaries to make black — meaning that we have subtracted all the reflected light (the paper looks black when no light at

ADDITIVE COLOR MIXING
(Lights shining on black background)

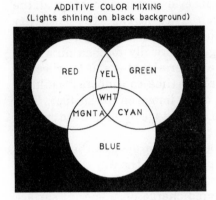

SUBTRACTIVE COLOR MIXING
(Paints on a white background)

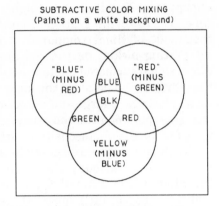

Figure 2.3   **Additive and subtractive color mixing**

all is being reflected).

There is a different system of primary colors that is used when we wish to create colors by mixing colored *lights*. This is the *additive* primary system, and those colors are red, green, and blue. If we mix equal parts of red, green, and blue lights, we will get white light. Note that two of the additive primary color names seem to be the same as two of the subtractive primaries — "red" and "blue." This is not the case — red and blue are the correct names for the additive primaries, but the subtractive paint primaries "red" and "blue" should technically be named, respectively, *magenta*, which is a red-blue color, and *cyan*, which is a blue-green color.

The relationship of these two systems of primary colors can be somewhat confusing. Figure 2.3 shows how overlapping circles of color interact for both additive and subtractive systems. The left part of Figure 2.3 shows that white light consists of an equal mixture of the three additive primaries — red, green, and blue — and the right part of the figure shows that an equal mixture of subtractive primaries produces black. The subtractive relationships are easily related to the additive situation by thinking in terms of what the subtractive primaries do to the *light* reflected off the white paper. The subtractive "blue" primary prevents reflection of red light, and it can therefore be called *minus red* — it filters out red light. If you look through a subtractive "blue" filter, anything red will appear black — it has been filtered out. Similarly, the subtractive "red" removes green light, and the subtractive

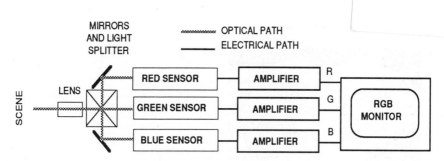

Figure 2.4   **Block diagram of a three-sensor RGB color video camera**

yellow removes blue light. Therefore, when we mix two subtractive colors, such as "blue" and yellow, we have removed both the red and the blue from the reflected light—leaving the green light—so mixing "blue" and yellow paint makes green. You can try the other combinations yourself to convince you that it agrees with what you learned in grade school.

Color Video

Let's return to the additive system, because that is the basis for color video systems. Video is an additive color system because the color cathode-ray tube (CRT) used for display creates three light sources which are mixed to reproduce an image. A color CRT mixes red, green, and blue (RGB) light to make its image. The colors are produced by three fluorescent phosphor coatings, which are on the faceplate of the CRT. Typically, these are scanned by three electron guns, which are arranged so that each of them impinges on only one of the phosphors. (There are many ways to do this.) The intensity of each of the guns is controlled by an electrical signal representing the amount of red, green, or blue light needed at each point of the picture as the scanning progresses.

So a color video camera needs to produce three video signals to control the three guns in a color CRT. A conceptually simple way to do this is to split the light coming into a color video camera into three paths, filter those paths to separate the light into red, green, and blue, and then use three video pickup devices to create the necessary three signals. In fact, many video cameras do exactly that, as shown in Figure 2.4. This kind of camera, known as a *three-tube* or *three-sensor* camera, is compli-

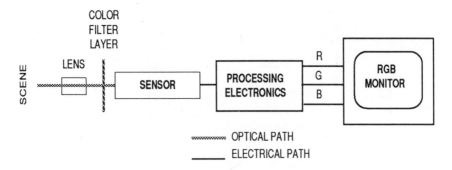

Figure 2.5   Single-sensor color video camera

cated and expensive because of the three parallel paths. A lot of the difficulty arises because the three sensors must be scanned in exact synchronism and exact physical relationship so that at any instant of time the three output signals represent the color values from exactly the same point in the image. This calls for extremely demanding electrical and mechanical accuracy and stability in a three-sensor camera design. The process for obtaining these exact relationships is known as *registration*. If a camera is not in exact registration, there will be color fringes around sharp edges in the reproduced picture. In spite of these difficulties, three-sensor cameras produce the highest quality images, and this approach is used for all the highest performance cameras.

There also are *single-sensor* color cameras, as shown in Figure 2.5. These use a system of filtering which splits the incoming light into spots of colored light which appear side by side on the surface of the sensor. When the sensor is scanned in the normal way, the electrical output for each spot in the image will consist of three values coming in sequence, representing the red, green, and blue values for that point. Because of the critical relationship required between the sensor and the color filter, it is customary to build the color filter right on top of the sensor's storage layer. Electronic circuits are then used to separate the sequential output from the sensor as it is scanned, into the required three separate signals. This approach is effective if the three spots of color can be small enough that they will not reduce the resolution of the final reproduction. Because that requires a threefold increase in the resolution of the sensor used (and that is difficult to come by), single-sensor cameras often are a compromise with respect to resolution. However, they are still the simplest, lowest cost, and most reliable cameras, and therefore single-

Home Video Cameras

sensor color cameras are widely used. All home video cameras are of the single-sensor type. Solid-state sensors are particularly suited to making single-sensor cameras. Because the resolution capability of solid-state sensors is steadily improving, single-sensor cameras are getting better.

Color Television Systems — Composite

The color cameras just described were producing three output signals — red, green, and blue. This signal combination is called *RGB*. Most uses of video involve more than a single camera connected to a single monitor, as we had in Figures 2.4 and 2.5. The signal probably has to be recorded; we may wish to combine the outputs of several cameras together in different ways, and almost always we will want to have more than one viewing monitor. Therefore, we will usually be concerned with a color video *system*, containing much more than cameras. In RGB systems, all parts of the system are interconnected with three parallel video cables, one for each of the color channels. However, because of the complexities of distributing three signals in exact synchronism and relationship, most color television systems do not handle RGB (except within cameras), but rather the camera signals are encoded into a *composite* format which may be distributed on a single cable. Such composite formats are used throughout television studios, for video recording, and for broadcasting. There are several different composite formats used in different countries around the world — NTSC, PAL, SECAM — and they will be covered specifically in the next section. Here we will concentrate on some of the conceptual aspects of composite color video systems.

Composite color systems were originally developed for broadcasting of color signals by a single television transmitter. However, it was soon found that the composite format was the best approach to use throughout the video system, so it is now conventional for the composite encoding to take place inside the camera box itself before any video signals are brought out. Except for purposes such as certain video manipulation processes, RGB signals do not exist in modern television plants.

All composite formats make use of the *luminance/chrominance* principle for their basic structure. This principle says that any color signal may be broken into two parts — *luminance*, which is a monochrome video signal which controls only the brightness (or luminance) of the image, and *chrominance*, which contains only the coloring information for the image. However, because a tri-stimulus color system requires three independent signals for complete representation of all colors, the chrominance signal is actually two signals — called *color differences*.

Luminance and chrominance are just one of the many possible combinations of three signals which could be used to transmit color information. They are obtained by a *linear matrix transformation* of the RGB signals created in the camera. The matrix transformation simply means that each of the luminance and chrominance signals is an additive (sometimes with negative coefficients) combination of the original RGB signals. In a linear transmission system there are an infinity of possible matrix transformations that might be used; we just need to be sure that we use the correct inverse transformation when we recover RGB signals to display on a color monitor. Psychovisual research (research into how images look to a human viewer) has shown that by carefully choosing an appropriate transformation, we can generate signals for transmission which will be affected by the limitations of transmission in ways that will not show (as much) in the reproduced picture.

In the color printing world there is another version of the luminance/chrominance system in use which has many similarities to that used in color television. That is called the *hue-saturation-value* (HSV) system or the *hue-saturation-intensity* (HSI) system. In this system, value or intensity is the same as luminance — it represents the black and white component of the image, and hue and saturation are the chrominance components. Hue refers to the color being displayed, and saturation describes how deep that color is. In a black and white image, saturation is zero (and hue is meaningless), and as the image becomes colored, saturation values increase. The same terms, hue and saturation, are used with the same meaning in color television.

In a composite system, the luminance and chrominance are combined by a scheme of *frequency-interleaving* in order to transmit them on a single channel. The luminance signal is transmitted as a normal monochrome signal on the cable or channel, and then the chrominance information is placed on a high-frequency *subcarrier* located near the top of the channel bandwidth. If this carrier frequency is correctly chosen, very little interference will occur between the two signals. This interleaving works because of two facts:

1.   The luminance channel is not very sensitive to interfering signals that come in near the high end of the channel bandwidth.

This is especially effective if the interfering signal has a frequency that is an odd multiple of half the line scanning rate. In this case, the interfering frequency has the opposite polarity on adjacent scanning lines, and visually the interference tends to cancel out. The selection of carrier frequency for the chrominance ensures this interlace condition.

2. The eye is much less sensitive to color edges than it is to luminance edges in the picture. This means that the bandwidth of the chrominance signals can be reduced without much visual loss of resolution. Bandwidth reductions of 2 to 4 for chrominance relative to luminance are appropriate.

Thus a composite system is able to transmit a color signal on a single channel which has the same bandwidth as each of the three RGB signals we started with. The transmission is not perfect, but it is good enough to be the basis of our worldwide television systems. This packing of the three RGB signals into the same bandwidth we once used for only one (black and white) signal may seem like we are getting something for nothing, but that's not true. What is really happening is that we are utilizing the spaces in the channel which are unused when transmitting only a single television signal, and we are also making compromises in the reproduction of the color information based on knowing what the viewer can see and what he or she cannot see in the final image. Together, these two features allowed the development of the color television system as we know it today.

Color Video Formats — NTSC

The NTSC color television system is the standard broadcasting system for North America, Japan, and a few other countries. NTSC is an acronym for National Television Systems Committee, a standardizing body which existed in the 1950s to choose a color television system for the United States. The NTSC system is a composite luminance/chrominance system as described above, and its key numbers are given in Appendix A. An important objective for the NTSC system was that it had to be compatible with the monochrome color system which was in place with millions of receivers long before color television began. This objective was met by making the luminance signal of NTSC be just the same as the previous monochrome standard — existing monochrome receivers see the luminance signal only. Furthermore, the color signal present at the top of the bandwidth does not show up very much on monochrome sets because of the same frequency-interleaving which reduces interference between luminance and chrominance.

In NTSC the luminance signal is called the Y signal, and the two chrominance signals are I and Q. I and Q stand for *in-phase* and *quadrature*, because they are two-phase amplitude-modulated on the color subcarrier signal (one at 0 degrees, and one at 90 degrees — quadrature). The color carrier frequency is 3.579545 MHz, which must be maintained very accurately. (The reasons for the funny number and the accuracy are beyond the scope of this discussion — see the references on Color Television if you are interested.) Appendix A gives the matrix transformation for making Y, I, and Q from RGB. As already explained, the I and Q color difference signals have reduced bandwidths. While the luminance can utilize the full 4.5 MHz bandwidth of a television channel, the I bandwidth is only 1.5 MHz, and the Q signal is chosen so that it can get away with only 0.5 MHz bandwidth! (In fact, pretty good results are obtained if both I and Q only use 0.5 MHz — most TV receivers and VCRs in the United States have 0.5 MHz bandwidth in both chrominance channels.)

When the I and Q signals are modulated onto the color subcarrier of the NTSC system, they result in a color subcarrier component whose amplitude represents the saturation values of the image, and the phase of the color subcarrier represents the hue values of the image. NTSC receivers usually have controls to adjust these two parameters in the decoding of the NTSC color signal. (On NTSC receivers, the hue control is often called *tint*, and the saturation control is called *color*.)

It should be pointed out that the NTSC system was designed to deliver satisfactory performance with the kind of signals created by television cameras looking at real scenes. Today, we also can generate video signals with computers, and there can be problems if a computer-generated signal does not follow the rules when it is expected to be passed through an NTSC system.

Color Video Formats — PAL and SECAM

The *PAL* and *SECAM* systems, which originated in Europe, are also luminance/chrominance systems. They differ from NTSC primarily in the way in which the chrominance signals are encoded. In PAL, chrominance is also transmitted on a two-phase amplitude-modulated subcarrier at the top of the system bandwidth, but it uses a more complex process called Phase Alternating Line (PAL), which allows both of the chrominance signals to have the same bandwidth (1.5 MHz). Because of the different bandwidths, a different set of chrominance components is chosen, called U and V instead of I and Q. In addition, PAL signals are more tolerant of certain distortions that can occur in transmission paths to affect the quality of the color reproduction. Appendix A gives numbers for the PAL system.

The SECAM system (*Sequentiel Couleur avec Memoire*), developed in France, uses an FM-modulated color subcarrier for the chrominance signals, transmitting one of the color difference signals on every other line, and the other color difference signal on alternate lines. Like the PAL system, SECAM is also more tolerant of transmission path distortions. See Appendix A for more information.

Most television production and broadcasting plants are concerned primarily with the composite standard of the country in which they are located. However, it is now common for programs to be produced for distribution to locations anywhere in the world, regardless of the different standards in different countries. This has led to the development of equipment for conversion between different composite standards. Such equipment does a remarkable job, but it is complex and expensive, and even so it has performance limitations.

There is an effort in standardizing circles to develop a worldwide program-production standard that is not necessarily the same as any of the composite standards, but would allow easy and nearly transparent conversion to any of the world's composite standards for local distribution. This goal will probably be achieved by the use of digital video technology.

Because the television standards in use are more than 30 years old, it is not surprising that the industry is also considering possible new standards for a much higher-performance television system, usually referred to as High Definition Television (HDTV). A finished HDTV standard is in the future, and is discussed in more detail in the last chapter of this book.

Video Performance Measurements

Analog television systems cause their own particular kinds of distortion to any signal passing through. Remember, analog systems are *never* perfect. Analog distortions also accumulate as additional circuits are added, and in a large system all the parts of the system must be of higher quality if the picture quality is to be maintained. A single component intended for a large system has to be so good that its defects become extremely difficult to observe when the component is tested by itself. However, when the component is used repeatedly in cascade in a large system, the accumulation of small distortions becomes significant. Many sophisticated techniques have been developed for performance measurement in analog television systems.

All analog video measurements depend on either looking at images on a picture monitor or making measurements of the video waveform with an oscilloscope or waveform monitor (which is just a special oscilloscope for television measurements). Because monitors also have their own distortions, looking at images on picture monitors tends to be suspect. Image-based measurements also involve judgment by the observer and therefore are subjective. On the other hand, oscilloscopic evaluation of waveforms can be more objective and therefore waveform-based approaches have been developed for measuring most parameters. However, picture monitors are good qualitative tools for performance evaluation, particularly when the characteristics of the monitor being used are well understood and the observer is skilled. Of course, the fundamental need for picture monitors in a television studio is artistic — there is no other way to determine that the correct scene is being captured with the composition and other features desired by the producer and director!

One measurement on video waveforms that must always be done is the measurement of video *levels*. The television system is designed to operate optimally when all video signals are kept to a particular amplitude value or level. If signals get too high, serious distortions will occur and display devices may become overloaded. Similarly, if levels are too low, images will be faded out, and the effect of noise in the system will become greater. Most video systems have a standard video voltage level (such as 1 volt peak-to-peak) for which they are designed, and all level-measuring equipment is calibrated for that level. To simplify going between different systems that have different actual voltage standards, most oscilloscopes and other level indicators are calibrated in *IRE levels*. (IRE is an acronym for *Institute of Radio Engineers*, one of the forerunners of today's worldwide electrical engineer's professional society, the *Institute of Electrical and Electronics Engineers*.) This refers to a

NORMAL

LOSS OF HIGH
FREQUENCIES

LOSS OF MID
FREQUENCIES

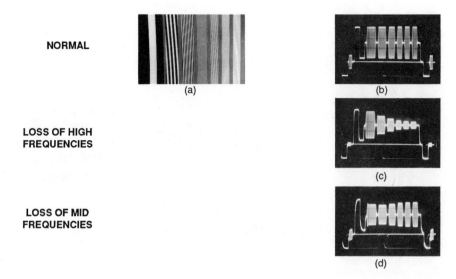

Figure 2.6   Testing frequency response with Multiburst pattern

standard for video waveforms which specifies that blanking level will be
defined as 0 IRE units and peak white will be 100 IRE units. Other
aspects of specific signals can then be defined in terms of this range.

Most video performance measurements are based on the use of *test
patterns*. These are specialized images which show up one or more
aspects of video performance. Test patterns may be charts which are
placed in front of a camera, or they may be artificially generated signals
which are introduced into the system after the camera. Because a
camera has its own kinds of impairment, a camera usually cannot
generate a signal good enough for testing the rest of the system.
Therefore, a camera is tested by itself with test charts, and then the rest
of the system is tested with theoretically perfect signals, which are
electronically generated by test signal generators.

Measurement of resolution
Let's begin with the characterization of resolution. One test pattern for
resolution is the *multiburst* pattern — it can be either artificially
generated or made into a test chart, and it is used to test horizontal
resolution of a system. A multiburst pattern is shown in Figure 2.6. It

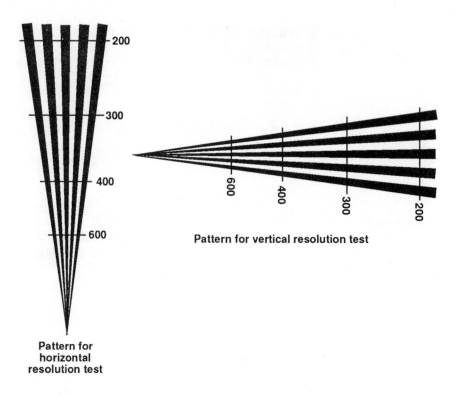

Pattern for vertical resolution test

Pattern for
horizontal
resolution test

Figure 2.7   Resolution wedge patterns

consists of sets of vertical lines with closer and closer spacing, which give
a signal with bursts of higher and higher frequency. Figure 2.6 also
shows a line waveform for correct reproduction of a multiburst pattern
and another waveform from a system which has poor high-frequency
response. This latter system would cause vertical lines to appear fuzzy
in an image. Another more subtle impairment is shown by the third
waveform in Figure 2.6 — in this case there is a mid-frequency distortion
which would make images appear smeared.

The multiburst pattern only tests horizontal resolution. To test
vertical resolution as well, a *resolution wedge* test pattern is used.
Figure 2.7 shows some resolution wedge patterns for testing both
horizontal and vertical resolution. A resolution wedge is used by
observing where the lines fade out as the wedge lines become closer
together. The fadeout line number can be estimated from the numbers

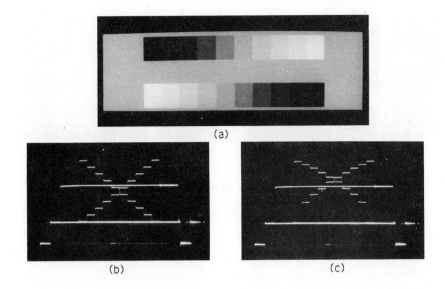

(a)

(b)   (c)

Figure 2.8   (a) Gray scale test chart, (b) normal waveform, (c) dis-
torted waveform showing black stretch.

beside the wedge — this would be the resolution performance of the
system or camera being tested. A common resolution test chart is the
EIA Resolution Test Chart, designed to test many parameters in addition
to resolution. It is usually placed in front of a camera. (EIA stands for
Electronic Industries Association, an industry group in the United
States which is very active in standards for television in that country.)
There are wedges for both horizontal and vertical resolution at different
locations in the image and various other blocks and circles to test camera
geometric distortion (*linearity*) and gray scale response.

Measurement of gray scale response
The gray scale test has its own special pattern, called the stairstep. This
may be either a camera chart or an electronic generator. Figure 2.8
shows one version of the gray scale chart, with examples of the signals
created by the pattern going through various systems. An important
consideration regarding gray scale reproduction in television systems is
that an equal-intensity-step gray scale test chart should not produce an

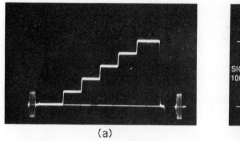
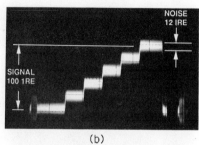

(a)   (b)

Figure 2.9   **Noise measurement using an oscilloscope.**
                 (a) Signal without noise. (b) Signal with noise.
                 S/N ratio =600/12=50:1=34 dB.

equal-step electrical signal. The reason for this is that the electrical signal will ultimately drive a CRT display, and the brightness vs. voltage characteristic of a CRT is not linear. Therefore, television cameras include *gamma correction* to modify the voltage transfer characteristic of the signal to compensate for an average CRT's brightness vs. voltage behavior. This is written into most television standards because the CRT was the only type of display which existed at the time of standardization. New-type displays such as LCDs may have different gamma characteristics for which they must include their own correction if they are to properly display standard television signals.

A typical CRT has a voltage-to-intensity characteristic close to a power of 2.2. To correct for this CRT gamma characteristic, a signal from a camera sensor having a linear characteristic (gamma of 1.0, which is typical for sensors) must undergo gamma correction by a circuit which takes the 2.2 root of its input signal. If gamma correction is left out of a system, the effect is that detail in the black regions of the picture will be lost, and pictures appear to have too much contrast regardless of how you adjust the display. In dealing with computer-generated images, it will be important to provide for the gamma characteristic to create realistic looking pictures on CRT displays.

Measurement of noise

Measurement of noise in a television system is an art in itself. Most

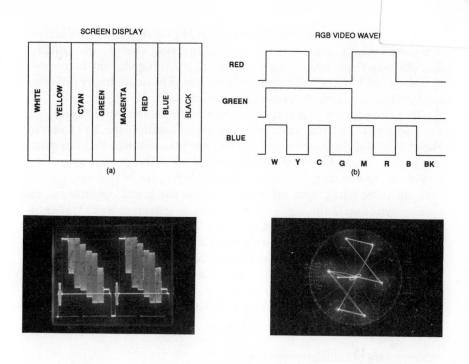

Figure 2.10   (a) Color bar test pattern, (b) RGB video waveforms,
(c) NTSC video waveform, (d) NTSC vectorscope display.

specifications are in terms of *signal-to-noise ratio* (S/N), which is defined
as the ratio between the peak-to-peak black-to-white signal and the *rms*
value (rms means *root-mean-square*, which is a kind of statistical
averaging) of any superimposed noise.

S/N numbers are commonly given in decibels (dB), which are logarithmic units specifically designed for expressing ratios. The *bel* represents a power ratio of 10:1, the *decibel* is 1/10th of that. Since signal-to-
noise ratios are usually voltage ratios, not power, a 10:1 signal-to-noise
ratio is 20 decibels because the power ratio goes as the square of voltage
ratio. Because the decibel is logarithmic, doubling the dB value is the
same as squaring the ratio: thus a 100:1 voltage ratio is 40 dB. A good
S/N ratio for a system would be around 200:1 (46 dB) which means that
the rms noise in that system is 200 times less than the maximum black-

to-white video level the system is designed to handle. Note that the measurement of rms noise requires an integrating kind of meter, and for this purpose there needs to be a region of the image for measurement that does not have any other signal present. There are many different kinds of test patterns for these measurements.

It is also possible to get an approximate S/N measurement by looking at video signals with a waveform monitor or oscilloscope, as shown by Figure 2.9. Noise appears on a video waveform as a fluctuating fine grain fuzz, which is usually evident on all parts of the active picture area of the waveform. (Noise usually does not appear during the blanking intervals because video equipment often regenerates the blanking interval, replacing sync and blanking — which may have gotten noisy in transmission or recording — with clean signals.) The peak-to-peak value of the noise fuzz can be estimated as a percentage of the total video black-to-white range. Then an approximate S/N ratio may be calculated by dividing 600 by the percentage of noise fuzz. This is based on the assumption that typical noise has an rms value six times less than its peak-to-peak value. Therefore, a 46-dB system would show about 3% peak-to-peak noise fuzz on its signals.

Measurement of color performance

Another class of measurement is involved with the color performance of a system. Measurement of color rendition of cameras is beyond our scope here, but there are simple means to test the rest of the system. This is most often done with a *color bar test pattern*. Figure 2.10 shows a color bar pattern and the resulting waveform display used widely in NTSC systems. The particular bar sequence chosen arranges the colors in order of decreasing luminance values, which makes an easy-to-remember waveform. The color bar signal can also be examined with a special oscilloscope made for studying the chrominance components — this instrument is called a *vectorscope*. Figure 2.10 also shows what an NTSC signal looks like on a vectorscope. There are also *split-field* versions of the color bar pattern, which put the bar pattern at the top of the screen and a different pattern (such as a stairstep pattern) at the bottom. With this arrangement, several different tests can be made using only one test signal. Available test signal generators provide many combinations of the test signals, to be used individually or simultaneously in split-field combinations.

There are many other kinds of video performance measurements, which are covered thoroughly in the references. For our purpose in understanding the interfaces between analog and digital video, the few which are covered in the foregoing paragraphs are sufficient.

Analog Video Artifacts

Archaeologists use the word *artifact* to refer to an unnatural object which has been found in nature—thus it is presumably manmade. In the video world, artifacts are un-natural things which may appear in reproduction of a natural image by an electronic system. In order to appreciate the vagaries of analog video, and particularly to appreciate the different things which an analog system does to video compared to what a digital system does, we need to give attention to these analog image distortions — artifacts. A skilled observer of video images can often recognize things that the casual viewer will never notice — but being able to recognize them is important when you are responsible for the system. Often the artifacts are clues to something that is starting to go wrong, recognizable before it becomes catastrophic. So, at the risk of destroying your future enjoyment of television images that may be less than perfect, let's look at analog video artifacts.

Noise

As we have already said, the most common form of noise is what we refer to on our television receiver as *snow*. The speckled appearance of snow is caused by excessive amounts of noise rather uniformly distributed over the bandwidth of the video signal, which is often called *white* noise or *flat* noise. Flat noise is readily observable when the signal-to-noise ratio falls below about 40 dB. A good three-sensor camera will generate a signal where the S/N is nearer to 50 dB. However, this signal may be subsequently degraded by various transmission paths or by video recording (see later discussion on recorders).

There are other types of noise which look quite different. For example, noise that is predominantly low-frequency (in the vicinity of the line frequency or lower) will appear as random horizontal streaks in the picture. The eye is quite sensitive to this kind of noise, but fortunately it is unusual in properly operating systems. When this kind of noise is seen, something in the system may have become intermittent or is about to fail catastrophically. Another noise phenomenon that looks different from snow is *color noise*. This is noise in the transmission path for a composite signal which appears in a frequency band close to the color subcarrier. Because of the relatively narrow bandwidth of the

Figure 2.11   Image showing analog smear artifact

chrominance channels, color noise appears as moderately large streaks of varying color. Color noise is primarily an artifact of video recording.

RF interference

Various kinds of coherent (not random) interferences can creep into video signals from other sources. The appearance of these interferences will depend on the exact frequency relationship they have to the scanning frequencies and to the color subcarrier. The degree to which they produce moving patterns depends on whether (or how closely) they are synchronized with the main signal. You have probably seen the interference which consists of two vertical bars spaced about 10% of the picture width which move slowly across the picture, taking several seconds to go all the way across. This is interference from another color television signal on the same standard but not synchronized with the main signal. It is quite common in signals received over the air, and it also may be a problem in a television studio system which has several

Figure 2.12   Image showing analog streaking artifact

sources of signals that are not synchronized.

Interference from single-frequency non-television signals will produce diagonal- or vertical-line patterns, either stationary or moving.
The relationship to the horizontal scanning frequency controls the exact
pattern produced. Interference from multifrequency sources will produce more complex patterns. For example, interference getting into
video from an audio signal will produce a pattern of horizontal bars
which changes in size and position with the sound. The relationship is
obvious if you are able to hear the sound while you watch the patterns
in the video.

Interference from coherent sources is much more visible than is
random interference or noise. This is because the coherent interference
creates some kind of pattern, which repeats over and over in the same (or
a slowly moving) location. Patterns of bars may have any spacing and
may range from vertical bar patterns through diagonal patterns to
horizontal bar patterns. Coherent interferences are often visible if they

exceed about 0.5% peak-to-peak relative to the black-to-white video range.

Loss of high frequencies
In a composite color television system, the first effect of loss of high frequencies is that the color saturation (vividness of color) will be reduced, or color may be lost entirely. (Except in the SECAM system: in SECAM the FM nature of the color subcarrier will retain color saturation. The effect becomes one of increasing color noise, or finally loss of color.) More severe loss of high frequencies will noticeably affect the sharpness of vertical edges in the image. In an RGB system, loss of high frequency will only affect sharpness, although if the loss is not the same in all three channels, it will also show as color fringes on vertical edges.

Smear
Smear shows up as picture information which is smeared to the right (usually). Figure 2.11 shows an image with smear. It is caused by a loss of amplitude or a phase shift at frequencies near or somewhat above the horizontal line frequency. Many tube-type cameras have a *high-peaker* adjustment, which can cause this kind of distortion if not properly set. It is also caused by long video cables which are not properly equalized.

Streaking
Streaking is present when a bright object in the image causes a shifting of brightness all the way across the image at the vertical location of the bright object. Figure 2.12 shows what it looks like. Streaking is usually caused by video information which gets into the blanking interval and then interferes with the level-setting circuits (called *clamps*) present in many pieces of video equipment. It may occur when a signal with an extreme case of smear passes through equipment containing clamps, or it can happen due to loss of frequency response at frequencies below the line frequency. The latter is usually caused by a component failure somewhere in the system.

Color fringing
Color fringing is present when edges in the picture have colors on them that were not in the original scene. This may be over all of the picture, or it may be confined to only one part of the picture. The principal cause of color fringing is registration errors in cameras. However, there are

some less frequent kinds of distortions in the frequency domain or in recording systems which will cause similar effects. In these cases the entire picture is usually affected, and it will most likely occur only on vertical edges.

Color balance errors

In an RGB system, the most likely color errors are those caused by the video levels not being the same in the three channels. This is a *color balance* error and it shows up as a constant color over the entire image. It is independent of which colors are in the scene. For example, if the red level is too low, white areas in the image will have a *minus red* cast — that is, they will appear cyan. All parts of the image will have cyan added to the correct color. Correct reproduction of white areas is a good test for color balance. In a composite system, color balance errors usually originate at the camera, where they may be caused by balance errors in the RGB circuits of the camera or errors in the composite encoder built into the camera. In NTSC and PAL systems, the subcarrier output goes to zero in white areas of the picture, so looking for zero subcarrier on a white bar is the test for proper balance of an NTSC or PAL signal. There are more subtle color balance errors where the color balance varies as the brightness level of the image changes. This is tested by observing the color balance on all the steps of a gray scale test pattern or test signal. The most common source of this kind of error is the camera.

Hue errors

In RGB systems, hue errors are unusual if the color balance is correct. However in composite systems, particularly NTSC, hue errors are caused by improper decoding of the color subcarrier; in particular, the *phase* of the color subcarrier controls hue in an NTSC system. NTSC television receivers have a hue control which adjusts this parameter. In a color origination system, however, the signals must all be set to a standard, so that the viewer will not feel that the hue control has to be readjusted every time a different camera is used. In the PAL system, the phase-alternating-line encoding approach makes the hue much less sensitive to the phase of the color subcarrier, and PAL receivers usually do not have a hue control.

Color saturation errors

It has already been explained that in composite systems, color saturation errors are most commonly caused by incorrect high-frequency response, which leads to incorrect color subcarrier amplitude.

Flag-waving

In low-cost video recorders, there can be a problem of synchronization instability at the top of the picture called *flag-waving*. It shows up as vertical edges at the top of the screen moving left to right from their correct position. In the recorder it is caused when the video playback head of the recorder has to leave the tape at one edge and come back onto the tape at the other edge (which happens during the vertical blanking interval, so this itself is not seen). If the tape tension is not correctly adjusted, flag-waving may appear. Some video recorders have a control called *skew*, which is an adjustment to minimize this effect.

Jitter

If the entire picture shows random motion from left to right (called *jitter*), the motion usually is caused by *time-base errors* from video recorders. In a video recorder the smoothness of the mechanical motion of the recorder's head drum is very critical in order to reproduce a stable picture. Higher-priced recorders contain a *time base corrector* (TBC) to correct this problem, but TBCs are expensive, and it costs less in a low-priced recorder to try to make the mechanical motion stable enough without a TBC. Occasionally, jitter effects will be caused by other kinds of defects or interferences getting into the synchronizing signals or circuits.

There are many more distortions and performance problems that can happen in analog television systems. If you really want to know more about this, consult the references.

Video Equipment

The video business is a mature, fully developed industry around the world. There are many manufacturers selling competitively to world-wide markets and because there are good standards for video signals and broadcasting, there is a wide range of equipment made for every imaginable purpose in video production, recording, postproduction, broadcasting, and distribution. The markets for video equipment range from the most sophisticated network or national broadcasting companies through industrial and educational markets, to the home (consumer) market. In the discussion that follows, we will simplify the market structure to just three categories:

1. Broadcast: Equipment used by large-market TV broadcast stations and networks. This equipment has the highest performance, is intended for large system application, and is the most expensive.

2. Professional: Equipment for use in educational and industrial applications (and smaller broadcasters) who cannot afford full broadcast-level equipment but still need a lot of performance and features.

3. Consumer: Home equipment where price is the first consideration. Must have simple operation and high reliability for use by a nontechnical user. The intended system application is very simple — usually one-camera, one-recorder systems.

Live-pickup color cameras

Broadcast-level video cameras for live pickup are generally of the three-sensor variety. The highest performance cameras are large units for studio use, and they have large-format sensors. They also support a very wide range of lenses for extreme zoom range, wide angle, very long telephoto, etc. These cameras can also be taken in the field when the highest possible performance is needed, but their large size hampers portability. There are smaller broadcast cameras, which are designed to be truly portable, and no expense is spared to keep the performance as high as possible. Broadcast cameras today generally contain computers which control their setup adjustments and many features of operation. However, the signal processing is primarily analog.

Professional-level video cameras are also of the three-sensor type, but they typically use smaller sensors and smaller optics to reduce the cost. They usually have a simpler system design with fewer special features; therefore, they are lower in cost and easier to operate and maintain. The compromises in picture quality and flexibility resulting from these changes have been chosen to be acceptable to the markets for this class of equipment. Applications are in education, training, and institutional uses.

Consumer-level video cameras go to the ultimate in cost reduction. They use the single-sensor format, which is a compromise in resolution performance compared to the other cameras, but it yields a small, reliable, low-cost unit. Consumer cameras are designed for high volume production and do not have a lot of special features. Very simple system application is intended. However, their performance and features have been good enough to create a mass consumer market, and they are an outstanding value in terms of what you get for the price.

Color cameras for pickup from film

Cameras specifically designed for television pickup from motion picture film or slides are called *telecine* cameras. Getting good television pictures from film is not as simple as it sounds because of two key problems. The first is that the frame rate for film is usually 24 frames per second, whereas television frame rates are 25 or 30 Hz. In the parts of the world where television frame rate is 25 Hz, film is shown on television simply by speeding the film frame rate up to match the television frame rate — a 4% increase. This amount of speedup is usually acceptable. However, for 30 Hz television systems, the 20% speedup required would be unacceptable. In these systems, the *3:2 pulldown* approach is used to resolve the different frame rates by the use of a special film drive mechanism. In the 3:2 approach, one film frame is scanned for three television fields and the next film frame is scanned for only two television fields. That is, two film frames are shown in five television fields, which is 2.5 television frames. This ratio of 2.5:2 is exactly the same as the 30:24 frequency ratio, so the average film frame rate can be the correct value. There is an artifact from 3:2 pulldown, which is a certain jerkiness in motion areas of the image — most television viewers in North America have become used to this effect and accept it.

The second film–television problem is a mismatch between film and television with regard to color reproduction capability. This arises because television is an additive color system, whereas film is a subtractive color system. Film images typically have more contrast than a television system can handle, and film shows its best colors in dark parts of the image (where the dye concentrations are the highest). Television gives its best colors in bright regions where the CRT is turned on fully and can overcome the effect of stray ambient light on the tube face. Both of these effects may be mostly overcome by using excess gamma correction (much more than is needed to correct for the CRT characteristic) to bring up the dark areas of the image from film. Telecine cameras have elaborate gamma circuits to provide this feature. In addition, since high gamma correction tends to also bring up sensor noise effects, telecine cameras need to start with a higher signal-to-noise ratio from the sensors in order to withstand the high gamma correction.

There are other problems of color rendition in reproducing film, which lead to a need in telecine cameras for much more flexible color adjustment circuits as well. It is common for telecine systems to contain very elaborate *color correctors* to deal with faded film, incorrect color balance, and other kinds of color errors. Because of the complexities of television from film, there is not much telecine equipment on the market outside of the broadcast field. Even in the broadcast field, telecine has become a specialized capability — used only for film-to-tape transfer. This is because video tape is much easier than film to deal with in a broadcast operation. Most film you see on television was transferred to tape some time before it was broadcast, often immediately after the film was processed.

Video recording equipment

For our use in digital video systems we will almost always be dealing with recorded video as the input to the digital system. Video will be shot with analog cameras, recorded, and often processed extensively before it is digitized. A whole industry, referred to as video *postproduction,* has grown up to take recorded video material and put it together into finished programs. Techniques and facilities for postproduction are highly developed and are serving large markets for television and other video production. We can expect that for some time these approaches will be the best way to create video for any use, including digital video.

Analog video recording is mostly based on magnetic technology. (An exception is the Laser Video Disc, which is optical.) Magnetic recording is not a good medium for use directly in analog recording, because it is highly nonlinear. Magnetic recording is in fact better as a digital medium, where a domain of magnetic material can be considered either magnetized or not magnetized. In video recording systems, this is dealt with by modulating the video signal onto an FM carrier before recording. The FM carrier is very well matched to the characteristics of magnetic recording, because it does not require sensitivity to different levels of magnetization; rather it depends on the size and location of magnetized regions.

One of the considerations of video recording for producing a program is that creation of the finished program requires going through the recording system several times. Original video is shot by using a camera with one recorder to get each of the scenes separately. A *SMPTE time code* signal is also recorded with all the material for use in controlling later processes. Recording scenes one at a time is done because it is much more efficient from the staging and talent viewpoint to not try to put scenes together in real time. This process of capturing the scenes one at a time is called *production*. To put together a program involving scenes from several cameras and often several locations, all the original tapes will be taken to a postproduction studio where the desired shots will be selected for assembly into the final program. Time code locations of all critical points will be tabulated. Then each of the scenes is run from its original tape under time code control and rerecorded in the proper sequence on a new tape to create an *edited master*. In that process various transition effects between scenes can also be introduced, such as dissolves, fades, wipes, etc. The edited master is usually backed up by making another edited master (called a protection master), or more commonly by rerecording copies from a single edited master (called a protection copy). Rerecording of video tape is called *dubbing*, and the resulting tape is called a *dub*.

If the dub from the edited master is the copy we use to digitize, you can see that we have gone through the analog recording process at least three times — once in original production, again in making the edited master, and a third time to back up the edited master. This is referred to as three *generations*. Since analog distortions will accumulate, a recording system that will deliver good pictures after three generations must have considerably higher performance than a recorder we will only go through once. In fact, three generations is almost a minimum number — there are often additional steps of video postproduction which can lead to needing five or six generations before the final copy is made.

Most video recorders are designed to record the composite video signal — NTSC, PAL, or SECAM. However, there are several new systems that use an approach called *component* recording. In a component recorder, the signal is recorded in two parallel channels, usually with luminance on one channel and chrominance on the other channel. Some of the recording artifacts can be reduced by this approach. If a component recorder is used with a composite camera, the composite signal must be decoded at the input of the recorder to create the component format. In this case, there is little performance advantage for a single generation because both component and composite signal degradations will be present. However, if the component recorder is either combined with a camera or used with a camera having component outputs, then the composite encoding does not have to occur until after recording, and the recorder performance can be improved. There are even some attempts to build component postproduction facilities where composite encoding does not occur until after postproduction. Such a facility is able to go through more generations and therefore can perform fancier postproduction effects.

Because of the need for multiple generations in video recording, there is a move in the television industry to develop *digital* video recorders. A digital recorder and a digital postproduction system could use unlimited generations, just as we are used to doing with computer recording devices. (In a computer we never worry about repeated loading and saving of data because we have confidence that the digital system will make no errors.) There are digital video recorders coming out in the broadcast field in both composite and component formats. The composite recorders use the same analog formats we have been talking about — they digitize the analog composite signal at their input. However, the digital component recorder takes a digitized YUV input format.

The standard of broadcast-level analog composite video recorders is the Type C recorder, using 1" video tape in a reel-to-reel format. This equipment is the workhorse of broadcast television around the world and delivers the best analog recording performance available today. The basic Type C machine is a large unit weighing somewhat more than 100 pounds and intended for fixed or transportable use. The Type C recorders will deliver good performance after three generations and are usable up to five or six generations. Another broadcast-level format, which is somewhat less used in the United States but is found extensively in Europe is the Type B system. The Type B recorders also use 1" tape, but their format is different in a way which allows smaller machines to be built. Type B performance is equivalent to Type C.

The two broadcast-level component recording formats, mentioned earlier, are the Betacam format and the Type MII format. Both of these use 1/2" tape in a cassette and have performance that is close to Type C level. Because of the small tape size, very compact machines can be built, including a camcorder format, which combines camera and VCR in one hand-held unit.

In professional-level recording, the 3/4" U-Matic format is the workhorse. This format uses a cassette with 3/4" tape, and machines come in rack-mounted, tabletop, and portable configurations. The 3/4" system uses a different way of getting the composite signal onto the tape, which requires that the luminance and chrominance be taken apart and then put back together inside the recorder. Doing that to a composite signal introduces some inherent degradations so that the picture quality of the 3/4" system is not as good as Type C, particularly with respect to color sharpness and luminance bandwidth. The 3/4" system typically can go only two generations with acceptable pictures.

In the consumer-level recording field we have the familiar VHS, Beta, and 8mm formats. These systems all use a method of recording similar to the 3/4" systems, in which separating of luminance and chrominance is required inside the recorder. However, because of the smaller tape sizes and lower tape speeds, bandwidths are much lower and the pictures are noticeably impaired by the recorder. However, they have proven themselves to be good enough for consumer entertainment use — witness the proliferation of consumer VCRs around the world. As a medium for input to a digital system, however, the consumer equipment is not very satisfactory. The reason is that these VCRs introduce artifacts that are different from the digital artifacts, and therefore when their signal is digitized by a low-cost digital system, both kinds of artifacts can appear. That is usually just too much! Note, however, that broadcast- and professional-level component systems (Betacam and Type MII) are based on the same consumer 1/2" tape technology, but that the component recording technique allows much better performance — at a higher price, of course. The component systems prove to be quite satisfactory for recording source material for digital systems.

Video monitoring equipment

Monitoring equipment for analog television includes picture monitors and waveform monitors, which also come in various price/performance levels. Broadcast-level video monitors cost several thousand dollars and come as close as possible to being *transparent*, which means that the picture you see depends on the signal and not the monitor. Broadcast monitors will also include various display modes which allow different aspects of the signal (in addition to the picture content) to be observed. One common feature is the *pulse-cross* display, which shows the synchronizing signal part of a composite video signal. Broadcast monitors are designed to be capable of being *matched*, so that a group of monitors will show the same signal the same way. Matching is important when several monitors are going to be used to set up signals that may eventually be combined into the same program. In such a case, it is important that the color reproduction of all signals match as closely as possible. Broadcast monitors always have inputs for composite signals. Some monitors also have RGB inputs, but this is unusual, because RGB signals seldom exist today in broadcast studios or postproduction facilities.

Professional-level monitoring equipment is designed to a slightly lower price/performance point — pictures are still very good but some features may be sacrificed in the interest of price.

Consumer-level monitoring is mostly done with the ubiquitous television receiver. All TV receivers have an antenna input for receiving the composite signal in RF form as it is broadcast on a TV channel. Recently, TV sets are also adding video inputs for a *baseband* composite signal, which may come from a consumer camera, VCR, or home computer. (A baseband composite signal is a video signal than has *not* been modulated up to a TV channel.) More rare is the RGB input, although this will appear in more TVs as the use of computers grows in the home. Technologies like DVI will eventually help that out!

Broadcasting of the composite video signal also introduces some characteristic performance problems, which we sometimes observe on our TV sets. In a wired system, it is unusual to get extremely noisy (snowy) signals or signals containing ghost images; however, these are common defects in over-the-air transmission. TV receivers are designed to deal (as well as possible) with these problems and still make an entertaining picture. Because broadcasting in a fixed channel bandwidth puts an absolute limit on the bandwidth for the video signal, pictures received on a TV will not look as good as they do in the studio. Also, there is a lot of competition in TV receiver manufacturing, and they are available at several points on the price/performance curve.

Summary

There are many other kinds of video equipment in use today for production and postproduction. We will discuss more of these later in Chapters 12 and 13 on digital motion video and special effects. The points to remember from this chapter are

- All video begins as analog (continuous values).
- Analog systems are never perfect.
- Analog distortions accumulate as you go through the system.
- You get what you pay for in video equipment.

And remember all the nomenclature introduced in italics — we will use it indiscriminately from now on. Now you are ready to take what you just learned and go digital!

3

Digital Video Fundamentals

A *digital video system* is any system where the information for images is represented by a collection of digital bits. The most common digital video system is the display portion of a personal computer. In a PC, the information for the video display is represented by a pattern of bits somewhere in the PC's memory, and the display is created by continuously reading from that area of memory, converting to a video signal, and displaying that on a raster-scanned CRT. This is usually done by special hardware in the PC, and it is done about 60 times per second so that the CRT display will not flicker. In the IBM PCAT and compatible computers, the above functions are performed by a plug-in card called a *video display adaptor*, of which there are several types.

However, our interest in this book is in *realistic* images, and most PC video display systems do not make particularly realistic images. Realistic images are created by analog video systems as we discussed in the previous chapter; this chapter will talk about how realistic images are converted to digital formats.

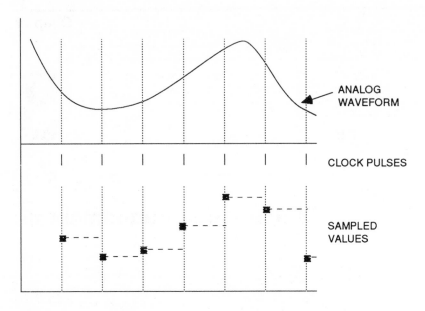

Figure 3.1   Sampling of an analog waveform

Sampling and Quantizing

The simplest (and the fastest) digital representation of numerical values is to use integer values, where we interpret a certain number of bits as powers of 2. This is widely used in computers, where a 16-bit digital word represents values from 0 to 2^{16} (that is from 0 to 65535), or (if the most significant bit is used for indicating a negative number) 16 bits represent values from -32768 to +32767. It is also convenient to use integer notation for the values needed to represent images; however, it is a problem (sometimes) that integer numbers inherently go in steps of 1, and fractional values do not exist. An integer digital number has a finite number of values equal to 2^n, where n is the number of bits used for the digital number. (Computers also commonly use *floating-point* numeric representation, which does not have the discrete step limitation of integers. However, floating-point requires much more difficult calculations in the computer and is not appropriate for image data except in certain non-real-time systems based on large computers.)

As we explained in Chapter 2, an analog waveform is continuous in

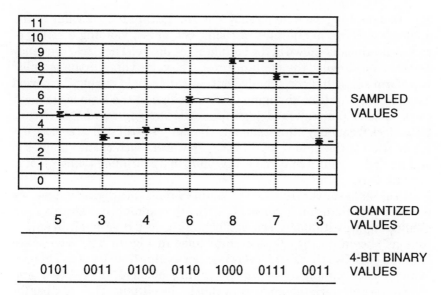

Figure 3.2   Quantizing

both value (amplitude) and in time.   In order to convert the analog waveform to a digital signal we must change both of these dimensions to noncontinuous values.   Amplitude will be represented by a digital integer having a specified number of bits, and the time will be represented as a series of those integer amplitude values taken at equal steps in time.   The process of converting time into discrete values is called *sampling,* and the analogous process for converting amplitude into discrete values is *quantizing.* These two processes together are referred to as *analog-to-digital conversion* (A/D conversion), or sometimes *digitizing.* Because we are changing the continuous-valued analog signal into discrete steps in A/D conversion, the process creates an approximation of the original signal.   However, as we will see, by proper choice of the numbers, A/D can be done with sufficient precision that approximation errors are invisible in the images reproduced by the digital system.   But most important — once the signal has become digital, no further approximations or errors need ever occur unless we want them to, because digital systems reproduce digital values exactly.

Figure 3.1 shows the process of sampling. As in most digital systems, we will establish a clock which has a frequency F_c and produces pulses at a period $P_c=1/F_c$. The clock pulses are shown in the center of Figure

3.1. In sampling, we take a reading of the instantaneous value of the analog waveform at the time of each clock pulse, yielding a string of sampled values as shown at the bottom of the figure. The clock frequency for sampling is called the *sampling rate*. The dotted lines in the sampled waveform show that the sampled output can be held during the period between clock pulses, changing to a new value at the time of the next clock pulse. The circuit for doing this is called a *sample and hold* circuit. If the clock frequency is high enough, the samples will still be a good representation of the analog waveform. However, each of the samples is still analog, in that it may have any value in a continuous range — it is discrete only in time.

In order to convert the analog sample values into digital values, we perform quantizing on them, as shown by the example in Figure 3.2. In quantizing, we establish a series of equally spaced thresholds in amplitude as shown by the 16 horizontal lines in Figure 3.2. The space between each two threshold levels is given a value from 0 to 15. Then we examine each sample to see which two thresholds it falls between and assign the appropriate numerical value to the output. This number for each sample is then represented digitally — a process called coding. The result of digitally coding a sampled signal is called a *pulse code modulated* (PCM) signal. Coding in general can replace the sampled and quantized values with any consistent set of numbers; it does not have to be a simple integer series. Such non-integer coding is called non-linear quantization or *companding,* and it is a technique used in some video compression schemes.

For our example with 16 quantizing levels, the numbers for the levels can be coded as digital integers with 4 bits, so we are quantizing at *4 bits per sample.* If we use enough quantizing levels (often a lot more than 16), the numbers will be an adequate representation of the analog signal, and the original signal can be reconstructed by an inverse process (which obviously would be called D/A conversion). However, before we look at D/A, we need to understand more completely the limitations of sampling and quantizing.

Limitations of sampling
We must have a high enough sampling rate and we must have enough quantizing levels — these are the main parameters of A/D conversion. Figure 3.3 shows how the sampling rate affects the results. In this figure, the analog waveform is a high frequency sine wave, which is sampled at three different rates. At the top, we have used a very high

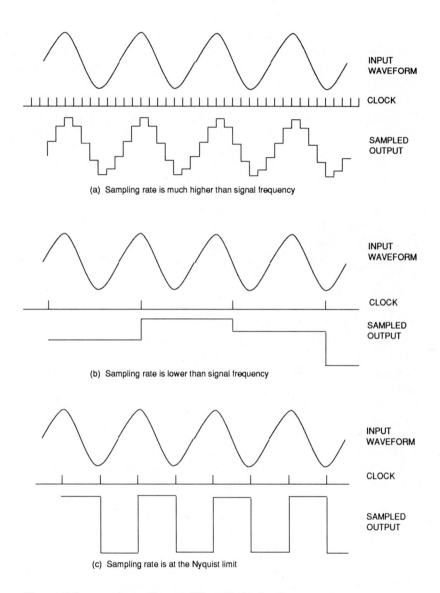

(a) Sampling rate is much higher than signal frequency

(b) Sampling rate is lower than signal frequency

(c) Sampling rate is at the Nyquist limit

Figure 3.3   Sampling at different clock rates

rate, so that there are nine samples taken for each cycle of the sine wave. It is clear that the samples in this case will fairly well represent the sine wave. The center example uses a sampling rate that is lower than the sine wave frequency, so that there is only one sample taken for every two cycles of the sine wave. It is also clear that the output of this sampler does not convey the sine wave at all. The example in Figure 3.3 is a special case where we are taking exactly two samples per cycle of the sine wave. In that case the sine frequency is present in the output, but the waveshape is square rather than sine. This special case is called the *Nyquist limit* — the input sine wave is at the highest frequency that can be reproduced with the sampling clock in the example. Turning that statement around — the sampling rate for a digitizing process must be at least *twice* the bandwidth of the signal to be digitized. If we try to digitize at a lower rate than that, the sampled output will be incorrect for any high frequency components in the input. Note that at exactly the Nyquist limit the frequency is reproduced correctly, but the output amplitude is ambiguous — consider what would happen if the input frequency is still sampled twice per cycle but at exactly the zero points on the sine wave — there is no output!

If we attempt to sample information above the Nyquist limit for the sampling rate we are using, spurious output components are produced. This is called aliasing — the output is false. More examples of aliasing will appear in the discussion of digital artifacts later in this chapter.

In order to avoid accidentally sampling input signals above the Nyquist limit and producing false output, it is good practice to include an analog low-pass filter on the input signal ahead of sampling in order to remove any input components above half the sampling rate. Most digitizing systems include such a filter. In a practical situation the sampling rate needs to be somewhat higher than twice the input bandwidth to allow the filter to function properly, since analog filters cannot be made to absolutely cut off above a certain frequency, rather they must cut off gradually with increasing frequency.

Limitations of quantizing

The limitation of the quantizing process is dependent on the number of quantizing levels which are used. However, unlike the very specific Nyquist limit, it is difficult to define an exact criterion for quantizing. Basically we must consider what the noise level in the input signal is (remember, all analog signals have noise) and choose only enough levels to reproduce signal variations that are larger than the noise — there is no need to have quantizing levels that would resolve amplitude changes smaller than the noise on the input signals. We will talk about this some more when we look at the A/D conversion processes for video and audio.

Video A/D Conversion

In most cases it will be desirable to have the digital video system use a component representation of the color signal. Either RGB, YUV, or YIQ formats are good possibilities. For reasons that will become clear when we look at what it takes to manipulate digital video signals, a composite color signal is not a good format for a digital video system. Another reason not to use a composite format is that a digital system can avoid the artifacts that come from composite encoding, as long as the input comes from an RGB signal that has never been encoded in a composite format.

Conversion of analog RGB signals to digital RGB format is done by applying the digitizing process described above individually to each of the RGB analog signals using three parallel A/D circuits. However, many times the input will have already been encoded into composite (particularly if it came from video tape). Conversion of a composite analog signal usually is done by analog *decoding* of the signal first into its YIQ, YUV, or RGB components, and then digitizing each one.

It is also possible to digitize the composite signal directly using one A/D converter with a clock high enough to leave the color subcarrier components undistorted and then perform decoding in the digital domain to get the desired RGB or YIQ format. Usually the clock is synchronized to the color subcarrier at a multiple of 3 or 4 times higher. The resulting digitized composite signal is decoded by digital processing; however, digital decoding in real time takes custom hardware in order to be fast enough to keep up with the signal. Chip sets for digital decoding are made for television receivers (some TV sets use digital processing of the analog signal to improve features and performance). These chips are designed first for low cost and they do not necessarily have all the features we might like, but their performance is consistent with almost all that you can get from the NTSC or PAL signal.

Pixels

The objective of video A/D conversion is to convert an analog signal that represents a real image into a digital signal representing the same real image. That is done as we described above by appropriately sampling and quantizing the analog signal to create a stream of digital numbers. That stream of numbers represents adjacent points in the image, following the same pattern in which the image was originally scanned by

(a) 512 x 480 pixels

(b) 128 x 120 pixels

(c) 64 x 60 pixels

Figure 3.4   An image digitized with different numbers of pixels

the camera. These points in the image are picture elements or *pixels*.
When reconstructing the image for display, each pixel will be shown as
a small rectangle filled with a color calculated from the number that
represents each pixel. Depending on the way the color is encoded in the
pixel values, and on the number of pixels used, the reconstruction can be
very realistic.

Any digital video display must be set up to have a specific number of
pixels in a two-dimensional array, horizontally and vertically. The
sampling clock used for digitizing must be accurately related to the
number of pixels desired and the scanning frequencies implied in the
input analog signal; otherwise the digital picture will have an incorrect
aspect ratio. Similarly, the number of pixels vertically must match the
number of lines in the analog scanning, or it can be an integer multiple
or submultiple. We will see later that there are image processing

PIXELS/HEIGHT

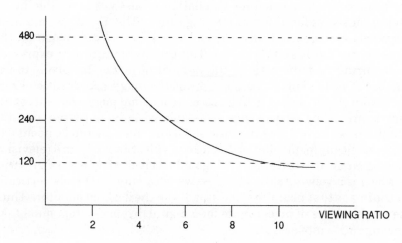

Figure 3.5   **Threshold of pixel visibility: pixel height vs. viewing ratio**

interpolation techniques which will let us have nonintegral relationships. These techniques will still require sampling initially with pixel counts that match the scanning, but then we can convert to different numbers of pixels with an interpolation technique.

Pixellation
The number of pixels in an image from a digital system is analogous to the horizontal and vertical resolutions of an analog system except that the appearance of pixels is much different from analog resolution artifacts. In an analog system with low resolution, the picture will become fuzzy, but in a digital system we get an effect known as *pixellation*. Figure 3.4 shows an image that has been digitized with different numbers of pixels. The blocky appearance of the images digitized with small numbers of pixels is the pixellation effect.

The degree of visibility of pixellation depends on how far you are from the image. In analog television this is spoken of in terms of the *viewing ratio* — the distance between the viewer and the screen as a multiple of the picture height. Viewing ratio is also a valid way to talk about the problem in digital video systems. Figure 3.5 shows the number of pixels per picture height needed for pixellation to just not be visible by an average observer for different viewing ratios. The figure assumes that the pixels are square, that is, they have the same dimension on the screen horizontally and vertically. This condition can also be expressed by requiring that the ratio of the number of horizontal pixels to the number of vertical pixels be the same as the image aspect ratio. From Figure 3.5 we can see for a 13 in. screen (diagonal measurement) at 4:3 aspect ratio being viewed from a distance of 18 in. (a normal desktop situation), we have a viewing ratio of 2.3 and there should be about 540 pixels vertically for pixellation not to be visible. Similarly, in a television viewing situation with a 25 in. diagonal receiver 9 feet across the room — which is a viewing ratio of 7.2 — we need only 170 pixels vertically for the image to appear the same as it did on the desktop monitor. Thus, the proper choice of pixel counts for a digital display system must take viewing ratios into account.

Removing blanking intervals

Digitizing an analog video signal with a constant sampling clock frequency will also digitize the blanking intervals. Normally there is no information of use to a digital system in the blanking intervals, so it is desirable to throw this information away and just make the digital signal out of the active picture parts of the analog input. This is easily done by storing the digitized information in successive locations in a digital memory and simply not storing anything during the blanking times. This will place successive scan lines next to each other in memory without any blanking at all. Of course, a computer will have no difficulty finding the start of a line in memory as long as it knows the number of pixels per line. It can multiply the pixels per line by the line number and add the starting address of the image to find the starting address of any line. Blanking or sync signals are not needed at all. Note that in this process we have placed a two-dimensional image into one-dimensional memory. (Of course, in many computer languages, a block of memory like this could also be defined as a two-dimensional array.)

Bitmaps

The arrangement of adjacent pixels and lines in a contiguous region of memory is called a *bitmap*. This name comes from the concept that we have mapped the physical image composed of pixels (which are made of bits) into adjacent addresses of memory. The bitmap is the only arrangement of pixel images in memory that will be used in this book. It should be pointed out that computer displays that are based on text characters (such as the IBM Monochrome Display Adaptor) do not use either pixels or bitmaps in memory. Text displays store an array of text characters, which are converted into a pixel format as they are being displayed — the pixels for the text image never exist in memory. (Note that some *compressed* image formats are also not bitmaps, and they may not have pixels either.)

Another image memory arrangement is the image plane or bit plane approach. In this approach, which is common in expensive workstations, there are a number of separate image memory arrays, each of which can store an image in bitmap format. The displayed image is made by combining the pixels from each of the image planes in real time as the pixels are read out from all of the planes in parallel. Combination of the plane pixels is done in hardware, and often there is a choice of the type of combination which can be used — for example, the pixel values may simply be added, which would give a mixture of the two images, or they might be *keyed* (logical AND) together, where the pixel value of one plane controls whether or not the pixel value of the other plane is displayed.

We have already defined two of the parameters of a bitmap — the starting address in memory and the length of a line in pixels. However, there are three other parameters needed to define a bitmap — *bits per pixel* (or *bpp*), *line pitch* (usually just called *pitch*), and the total number of lines (or pixels in the vertical direction).

Bitmap pitch

The pitch value for a bitmap specifies the distance in memory from the start of one line to the start of the next (adjacent in the image) line. Pitch can be specified either in bytes or in pixels. In this book pitch will always be given in pixels — thus, the actual address shift in bytes of memory from one line to the next is equal to pitch multiplied by bits per pixel/8. If the bitmap is packed as was described above (with no blanking intervals), then the pitch is equal to the pixels/line. However, the pitch can be a larger value — this is sometimes convenient even though it may seem to be wasting memory. In some configurations, other information can be placed between the lines — even another bitmap with the same pitch but a different starting address. The most common use of pitch different from pixels/line is to specify pitch to the next highest power of 2, which will help certain kinds of programs to run faster. (More about that later.)

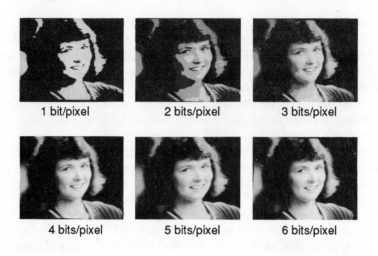

1 bit/pixel   2 bits/pixel   3 bits/pixel

4 bits/pixel   5 bits/pixel   6 bits/pixel

Figure 3.6   Images at different bits/pixel showing contouring

Dealing with interlaced scanning

When we are using standard television signals as the input for digitizing, we must decide what to do about the interlaced scanning — we only get half of the lines from each vertical scan. In order to make various digital processes easier, it is desirable that the bitmap lines should not be interlaced in memory, but rather we should build a non-interlaced bitmap. This is handled simply by using double the line pitch while writing into memory during digitizing, so that the lines from each field of the input signal will be placed alternately in memory after two complete vertical scans have been digitized.

Bits per pixel

The bits-per-pixel value determines how many colors the video system can display simultaneously. Different systems use different values, ranging from 1 to 32 bits per pixel. A 1-bit-per-pixel system can only display two colors, usually black and white, black and green (with a

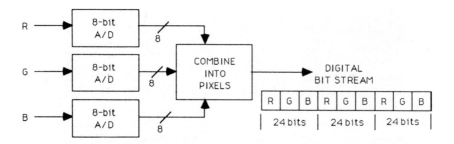

Figure 3.7   Analog/digital conversion with 24 bpp output

green monitor), etc. However, this arrangement takes the least possible amount of memory to hold a bitmap image. There is no gray scale, nor is it possible to have other colors at the same time.

A 1-bit-per-pixel image with no gray scale and no color distinctions cannot be called a realistic image. Although such images have many uses in the computer graphics world, they are nowhere near our objective in this book to talk about reproducing realistic images. What does it take to reproduce a truly realistic image? Let's begin discussion of that by looking at monochrome.

Remember, we said in Chapter 2 that a good signal-to-noise ratio for a video system was 46 dB, which would produce a video signal with about 3% peak-to-peak noise on it. Taking the criterion outlined above for the number of discrete levels required to reproduce a signal with noise, we might choose to set one level equal to the peak-to-peak noise — that would require 33 levels for our 46 dB system with 3% noise. This could then be reproduced by digitizing with 5 bits (32 levels). However, when we look at pictures that are quantized at 5 bits, we will probably not be happy because there will be a digital artifact known as contouring, which is clearly visible. Figure 3.6 shows the degree of contouring caused by digitizing at bits-per-pixel values ranging from 1 to 6.

Contouring

Contouring occurs because all analog levels which fall between two thresholds of the quantizer are replaced with the same digital value, and they will thus be reproduced at the output of the digital system with the same value also. Any part of the image that has a slowly changing analog level will be converted into stairsteps by the quantizing. The stairsteps will be clearly visible even when they are smaller than the random noise that was on the analog signal because they are correlated with the image itself. (The analog noise was random and constantly changing — that is why 3% noise is not very visible.)

For noise patterns that are coherent, or correlated with the image, the visibility threshold in luminance level is more like 0.5%. That is 1/200 of the amplitude range, and would compare to quantizing with 8 bits per pixel (256 levels). For contouring to be invisible in monochrome or luminance signals, 8 bits-per-pixel quantizing is required. There is not much advantage to using more than 8 bits on a signal which will be displayed visually, unless a nonlinear process such as gamma correction is to be done after digitizing. (In some cases gamma might be done other than in a camera, even after digitizing, in order to create some kind of special effect.) Additional bits might also be required if the signal is going to be analyzed in some way other than visually.

Color bits per pixel

All the foregoing discussion has been about monochrome images. What about color? Obviously we have to perform similar processes on each of the three color components (RGB, YIQ, YUV, etc.) to reproduce color images digitally. Normally this is done as shown in Figure 3.7 with three A/D circuits operating in parallel on the three component signals, and then the results are combined on a pixel basis to create a series of color pixels. The simplest arrangement, which applies to RGB systems, is to do 8-bit conversion on each of the color signals, and then combine the digital bits into 24-bit pixels, 8 bits for each primary color. This system gives excellent reproduction; the 24 bits are capable of conveying more than 16 million different colors! (2^{24}=16,777,216) However, 24-bit systems are expensive in terms of memory usage, storage, and processing power needed, so many techniques have been developed to use fewer bits per pixel and still have satisfactory color. For reference, a single 512x480 pixel, 24 bpp image will require 720 kilobytes of storage.

Another digital format which we have already mentioned is to A/D convert an NTSC or PAL composite signal and handle that digitally. 8 bpp turns out to be good enough for that, but the sampling clock must be at least 3 times the color subcarrier frequency — this leads to an image size of about 350 kilobytes. This is only a small improvement from the 24 bpp component system, and considering the performance compromises of the composite signal to begin with and the extreme difficulty of performing any processing on the image, digitizing of composite signals is not a good choice for computer video systems. (A new series of broadcast digital video recorders — the D-2 standard — is now coming on the market, and they are based on digitizing the composite television signal. This is appropriate for the television business because digital recording overcomes the analog limitations on multiple-generation recording, and the broadcast television system uses the composite signal anyway. Furthermore, processing of the signals, which would be difficult in the composite-digital format, can be done in the analog system before or after the digital recorder.)

It is possible to create quite realistic images using 16 bits per pixel by a compromise in the number of bits allocated for each color. In an RGB system, the values can be truncated to 5 bpp for each color and then assembled into 16 bpp words (one bit left over). This system is capable of 32,768 colors and works well with highly colored images. Its weakness (contouring) will show up in images which have subtle shadings of colors over large areas.

Techniques to reproduce good images with fewer bits than the above numbers are called video *compression* techniques. There are many possibilities, which will be covered in Chapter 11. However, it should be understood that the use of video compression implies many things other than simply the image quality compromises that may be necessary. Compression requires some processing both for the compression at the input and the decompression at the output. Since we are dealing with a computer here, we could expect that our CPU will do any processing necessary — it's certainly capable of doing anything we want to program it for. However, we will quickly come up against speed issues — video compression and decompression techniques are real hogs of CPU cycles, and things will become much too slow if we attempt anything fancy. This leads to special hardware being used for most compression-decompression systems, with its obvious effect on cost and complexity. The second point about compression is that it quickly becomes awkward to perform any manipulation on images while they are in the compressed form. We must first decompress to an easily manipulated format like RGB, and then recompress after the manipulation. Later we will discuss how the newer systems deal with these difficulties of compressed video systems.

Color-Mapping

There is another way of using a bitmap to represent an image where the pixel values do not directly represent the color sample values, but rather the pixel values are an index into a *color look-up table* (CLUT) which holds the actual color value. This technique, which is called *color-mapping*, allows the number of colors represented by the pixel value to be a subset of a much larger set of colors (the *palette*) represented by the bits-per-pixel in the color table. This subset of colors contained in the CLUT may be changed for each image displayed, or in some systems it can even be changed for different lines of the image while the image is being displayed. For example, we might have 8 bpp in the bitmap but 24 bpp in the CLUT. This would be a system which can display 256 colors out of a palette of 16,777,216 colors.

To display a realistic image with a CLUT system as described above, the image must be processed by an algorithm which chooses the 256 colors that occur the most to create the CLUT. Then the pixels are assigned the index value from the table for the color closest to each pixel's actual color. Both the CLUT and the pixel values would be then stored to save the image. When the image is loaded for display, the CLUT is loaded into the hardware for that, and the pixels are loaded into the display memory. When the display hardware reads out the pixels to refresh the display, the pixel values will be converted to the color values by looking them up in the CLUT in real time.

Color-mapping with a modest number of colors such as 256 does not usually create very realistic images — images with 256 or fewer colors may have a "cartoon" appearance. If you tailor the color map to the image (which may be an expensive process), you can often get fairly good realistic images with only 256 colors. However, if you compare the reproduced image to the original, there will be quite noticeable wrong color errors; but the picture will still be quite acceptable when the original is not available for comparison. An important exception to this is skin tones, which may require too many of the CLUT colors to obtain satisfactory smooth shading.

Color-mapping is still probably the simplest approach to compression down to 8 bpp that delivers an approximation to realism. Obviously the results will depend on the number of colors present in the input image — images with few colors may look good whereas elaborately colored or shaded images may show severe contouring. In Chapter 11 we will discuss more sophisticated compression approaches which can get a lot more out of 8 bpp. However, there is a place for color-mapping in a digital video system, because it is a format which provides a lot of computer graphics capability without the need for much real-time

Table 3.1 — Video Display Adaptors for PC AT

Typical Board	Pixel Resolution	No. of Colors	Memory bytes for 1 image	Approx. Price
CGA	320x200	4	16,000	$75
EGA	640x350	16	112,000	$200
VGA	640x480	16	153,600	$300
Targa 16	512x480	32,768	491,000	$2000
Targa 32	512x480	16,777,216	983,000	$5000

processing. Therefore, most systems will contain options or modes for CLUT use.

This is a good point at which to divert a little and show the relationship between the realistic video systems we have been discussing and the common video display formats now used with computers. Table 3.1 lists some various types of display adaptors for the IBM PC AT and compatible computers along with some of their characteristics given in the terms we have defined above. Note that only the Targa boards are capable of realistic images. (The Targa boards are products of Truevision, Inc. They are described in more detail in Chapter 10.) Note also the large difference in memory requirements of these boards compared to the others. Many of these boards also offer different combinations of resolution and number of colors. Table 3.1 shows the formats having the largest number of colors.

In many digital video systems there will be a need to allocate some of the bits in a pixel for purposes other than color values. For example, it may be desirable to define a region of the image where another image (maybe from an outside source) will be placed (or *keyed*) into the first image. With appropriate hardware responding to one bit in each pixel, this technique can provide a window of any shape on the screen where video from an external VCR or other live source could be displayed. Of course there must be proper synchronization of the signals (called *genlocking*) for this to work correctly. This technique can be carried further to use 8 bits from each pixel to control a video mixer, which will allow a video dissolve between the computer's image and an external source. 24-bpp systems are often extended to 32 bpp for such purposes. In most cases, application of extra bits like this will require additional special hardware either inside the system or externally to respond to the extra bit values and perform the desired process.

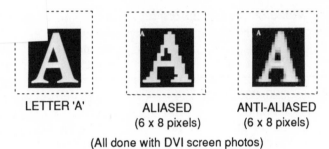

LETTER 'A'   ALIASED   ANTI-ALIASED
                   (6 x 8 pixels)   (6 x 8 pixels)

(All done with DVI screen photos)

Figure 3.8   **Text character "A" reproduced (a) at high resolution,
                   (b) at 1 bpp with 16x20, (c) anti-aliased with 8 bpp**

D/A Conversion

Realistic images may be displayed on an analog RGB monitor. Therefore, to drive such a monitor from our digital display refresh circuits, we
must convert the digital data in real time back to analog values. This is
done with a *D/A converter*, which is typically implemented on a chip
(sometimes three to a chip for RGB). The D/A converter simply contains
some digital switches, which are controlled by the signal bits to deliver
analog values equal to the value represented by each bit. These values
are summed at the output to create an analog signal. Usually the D/A
will also include hold circuits, which ensure that the output value will
change smoothly and monotonically from one pixel to the next.

The visibility of pixellation can be somewhat reduced by including an
analog filter on the output of a D/A converter. A filter with a bandwidth
equal to the Nyquist limit for the pixel rate used will not change the
resolution very much, but it will smooth some of the sharp transitions
between adjacent pixels — but only in the horizontal direction. Not all
digital video systems use these filters, however. Of course the results
will be best when there are enough pixels that the eye will not see them
anyway — then the question of filtering becomes moot.

Digital Video Artifacts

We have already mentioned the most important digital artifacts —

pixellation, contouring, and aliasing. Figure 3.4 shows the effect of pixellation and Figure 3.6 shows contouring. Aliasing is the error caused by sampling an image that contains frequency components (or resolution detail) above the Nyquist limit for the sampling rate. A classic example is shown by the reproduction of text characters in the typical computer display. A theoretical text character contains infinite resolution; however, to display a character on a computer screen with finite numbers of pixels, we perform a sampling process to create the character image in pixels. Figure 3.8 shows this for the text character "A," being sampled with an array of 16x20 pixels. The jagged effect on the diagonal parts of the "A" is the result of aliasing. If the text is reproduced without gray scale, there is no way around this. However, if the system has gray scale capability, then an *anti-aliasing* process can be done on the character to fill in some of the jagged spots with intermediate gray levels. (Figure 3.8 shows the character very large so you can clearly see the effects of aliasing and anti-aliasing. To get a better idea of how this would actually look on a computer screen at the proper viewing distance, figure 3.8 should be viewed from a distance of six feet or more.) As shown in Figure 3.8, anti-aliasing is an effective way to reproduce text when you have multiple bits per pixel. In fact, a television camera looking at the original text character will perform a similar process because of the finite size of its scanning spot and the continuous nature of the camera output signal.

Summary

We will be taking the subject of digital video much further in the later chapters of this book. The key points to remember here are

- The limits of digital video system performance are determined by the choice of number of pixels and bits-per-pixel in the original A/D conversion.
- Realistic images require large numbers of pixels and particularly large numbers of bits-per-pixel compared to computer graphic applications.
- Once the above parameters are chosen, there are many possibilities for how images can be handled in subsequent digital processing, but one must be careful to not further compromise picture quality.

4

Digital Audio Fundamentals

You may ask "What is audio doing in a book on digital *video?*" The answer is in the word *realistic*, which we have used so much already. Realistic video isn't very effective unless it is also accompanied by realistic sound. Turn off the sound on your television and see how much you get out of that interview program or the evening news show. And what would a music video be without the music? Sound is an integral part of realistic video. Of course we could have the computer create synthetic sounds, but that too is a gross mismatch with a realistic picture. Therefore, we need to look at techniques to bring real sound into the computer just like we have reviewed the technology for bringing real video to the computer screen.

You might also think that digital audio would be simple compared to digital video since the bandwidth of audio is hundreds of times less than the bandwidth for video. It isn't so! Digital audio calls for many of the same considerations as digital video, but there's at least one aspect which makes audio more difficult than video. We cannot get away with imposing a structure on audio like we can when we scan the video. The concept of lines and frames do not fit audio — audio has to be a continuous signal with no blanking intervals or other structural features that could make the signal easier to handle. Furthermore, there's no such thing as a *still frame* in audio. In video, if things get ahead of us, we can always display a still frame while the processing catches up — and if it's not too long, the viewer may not even notice. In audio we can never stop — to do so will cause an abrupt loss of information, usually a loud sound, and a loss of synchronism (or *lip-sync*) with the video. So audio has its own kinds of problems which more than compensate for its lower bandwidth.

There are some applications for realistic audio without realistic video. Real audio is so effective — and so unexpected with a computer — that it will find use as a help medium, as a background environment medium (music, motor sounds, etc.), or simply as a feature in itself such as using it for a telephone receiving and answering medium.

Just as we did with video, we will begin our discussion of digital audio by first looking at the analog world. Natural sounds are analog — minute pressure variations in the atmosphere which we can pick up with various kinds of pressure-to-electrical transducers. Of course such a transducer made specifically for sound pickup is a *microphone*. It is tailored to the frequency range and the pressure values which relate to the audibility characteristics of the human ear. The microphone produces an analog electrical signal which is as closely as possible a replica of the pressure variations in the air surrounding the microphone.

Audio Production Equipment

A full range of equipment is available for storing, processing, and reproducing analog audio signals. Even more so than video, audio technology is mature and available for a wide range of applications and price/performance levels. We will not go into full detail here, but audio equipment can be categorized into the same groupings we used for video — broadcast, professional, and consumer. You are probably already familiar with much of the equipment available. The audio production industry has many clients — more so than video. There are the audio record, tape, and compact disc producers; the film makers; the television business; education; and industry. All of these are looking for much the same service from audio production equipment.

Audio production is done with microphones and tape recorders.

Audio recorders for production usually have only one or two tracks on magnetic tape. Video recorders also record audio of course, but in many broadcast-level or professional-level productions, the audio will be recorded separately and go through separate postproduction. Only at the end of the process will the audio and video be combined on the same tape. This means that audio production also needs to work with SMPTE time code for proper synchronization with the video.

There are large facilities available for audio postproduction in most major cities around the world. In these studios, audio from many sources may be equalized, effects may be inserted, and the sources combined (*mixed*) into the tracks which will be used by the final application. Typically an audio postproduction house will work with many parallel tracks on magnetic tape — 24-track machines using 2 in. wide tape are common. Separate mix tracks for music, dialogue, and effects will be developed, and a final mix of these to the output tracks will be done. If the output tracks are in stereo, positioning of sound sources in the stereo field will be accomplished during the various mixes. Audio postproduction is an art in itself and commands a lot of the attention of a producer because the audio is a crucial part in conveying the message intended by the audio-video material being produced.

Just as in the broadcast-level video business, there is a trend in audio production and postproduction to go to digital equipment. Products for digital recording of audio are on the market at broadcast-level performance and prices — the motivation is the same as for video — unlimited generations of recording. In the audio business there is already a digital medium for distribution of audio to consumers — the Compact Disc — and this also has been a major motivation for the audio production industry to convert everything to digital.

For our interest here in bringing real sounds into the personal computer environment, we first must convert the analog audio to digital using an A/D converter (unless we have captured the audio on one of the new digital tape systems — then we only face a format conversion process). The A/D process is exactly the same as we discussed for video — sampling and quantizing; however, the frequencies and the precision are quite different. For audio, where the maximum bandwidth is 20 kHz, sampling rates need to be around 44–50 kHz. This is slow compared to the MHz rates required by video, but that is partially offset by the need to have many more bits per sample. In video, we talked about 46 dB signal-to-noise ratio as being good, but that number is terrible for audio. Audio S/N ratio needs to be above 70 dB for noise not to be heard in normal listening, and when the considerations of the production process are included, S/N ratios in the 90 dB range become necessary. Therefore the standard of audio digitizing is to use 16 bits per sample. This gives a theoretical S/N ratio of about 96 dB.

Audio Compression

Quantizing of 16-bit digital audio is normally done with *linear* PCM, where each step of quantization has equal size. This is the format used with the audio compact disc. However, with fewer than 16 bits per sample, other strategies are worthwhile. Nonlinear PCM or *companding* can be used where the steps of quantization at low signal levels are smaller than the steps at high signal levels. This tends to reduce the *quantization noise*, which otherwise may be heard when the sound level is low. Quantization noise occurs at low signal levels because the quantization steps become a large fraction of the signal amplitude. The companding strategy is useful in the range of 12–14 bits per sample.

For even further compression of audio one can use *differential* PCM or DPCM. In this scheme, the quantizer codes the *difference* between adjacent samples instead of the samples themselves. Since most of the time the difference between samples will not be as large as the samples themselves, fewer bits can be used for coding. If adjacent audio samples tend to be similar in amplitude then DPCM will work well. However, this will not be the case when the audio contains a lot of high-frequency information which is getting near the Nyquist limit. Then adjacent samples will be very different, and a DPCM system will not be able to keep up. The result will be severe distortion heard on loud high-frequency passages of the sound.

A further improvement is to make the DPCM system *adaptive*. In this system, called ADPCM, the scale factor of the difference bits is changed by some strategy which is sensitive to the signal level or degree of high-frequency content. This means that when the signal is small, the difference bits will control small steps of amplitude, but if the signal becomes large, the range of voltage represented by the difference bits will be cranked up. The information for control of the scale factor is also contained in the stream of difference bits, so that the decoder for ADPCM can find out how to control the scale factor when playing back the sound. There are many different systems of ADPCM, but the general idea of all of them is as described above. Dedicated integrated circuits are available to do some of the ADPCM algorithms, particularly one which has been standardized in the telephone industry by the *CCITT* (Consultative Committee for International Telephone and Telegraph). This algorithm is intended for telephone speech and operates at 32 kbits per second, using ADPCM at 4 bits per sample and with a sampling rate of 8,000 per second. It provides very satisfactory speech at telephone quality levels, and is suitable for use on personal computers. Higher quality sound can be achieved with a similar 4-bit-per-sample ADPCM system simply by increasing the sampling rate and therefore the system bandwidth.

ADPCM systems break down primarily with regard to the adaptive behavior. When a large amount of adaptation is required by a sound which changes too fast for the system, distortion may be heard. The degree of audio compression obtained is limited by those kinds of artifacts, and audio compression of more than 4:1 is difficult to obtain. This is a direct result of the lack of structure in an audio signal — structure offers possibility for redundancy in a signal, and removal of redundancy offers opportunity for compression. We will see later that much larger compression factors can be achieved with video.

Audio Is Real Time

We have already said that audio must play continuously — it cannot successfully be interrupted. To a computer, this means that audio is a real-time operation and it must have a high priority compared to other things going on in the computer. Many personal computer operating systems (such as MS-DOS) do not provide for real-time operations, and therefore audio on these computers must have a path around the operating system. This can be done by using the interrupt capability of a PC AT, except that there are DOS functions which disable the interrupts, such as disk access. In a PC AT we would like to use the hard disk to store audio, and while playing we will need to use DOS to get the data and pass it to the audio hardware. To keep the audio playing while more data is being fetched usually requires that the audio hardware contains *buffering* on its data input.

Buffering in the audio hardware will allow us to fetch audio data in blocks large enough that there will always be audio data stored ahead in the audio hardware, therefore we will not run out of data during the longest possible disk access. Usually the rest of the application (video and computer functions) can be designed so that an access longer than several seconds is never required without an interrupt. (That means if longer accesses are required, they must be called in pieces by the application to allow interrupts to get in.) Since the audio data rate on the average is quite low (4,000 to maybe 16,000 bytes per second), and it's not too much to ask the audio hardware to have 16 kbytes of memory on board, we can buffer up to four seconds of audio to cover disk accesses (the capability to do buffering is a key advantage of digital systems over analog systems).

One disadvantage to buffering is that it implies some delay in starting because we must fill up the buffer before we begin playing audio. There also can be a delay in making a change to any audio that is playing if the change also has to go through the buffering. A good system will have a strategy for getting around that, so that the audio hardware can respond instantly when necessary. In Chapter 8 we will be talking in more detail about a specific audio system that has all these features.

Synthetic Digital Audio and the Music Industry

A tremendous amount of technology has been developed for creating synthetic sounds digitally. The main thrust of this work has been the music equipment business; in fact that field is a place where the personal computer and audio have already joined forces, and therefore it is relevant to our interests here. Electronic musical instruments began as keyboard devices using analog technology. These instruments were known as *synthesizers*, which combined a keyboard controller with analog sound generators. As digital technology became available, synthesizers became digital. Today, however, the controller and the sound generation have become separated, and we have digital sound generators which can be controlled by any kind of player interface. If you are a wind instrumentalist you can get an electronic controller which is played the same as your clarinet, or if you prefer to play guitar, there are electronic controllers which play like guitars — in either case, the sound generation is electronic.

Music Instrument Digital Interface
The key to this flexibility is the *Music Instrument Digital Interface* (MIDI), which is a digital bus specification for connecting music controllers and musical devices. Essentially all professional electronic music equipment today has a MIDI interface on it. There are keyboards which

do not do anything except send out MIDI signals as you play, and there are music modules which can respond to MIDI signals and make almost any kind of sound imaginable. There also are instruments which combine controller and sound generation, but even these have MIDI so that they may be combined with other instruments to produce *layers* of sound.

One kind of sound generator that is gaining acceptance in the music industry is the *sampler*. This instrument samples and stores real sounds and then can play them back with pitch change. For example, in the simplest case you can sample one note from a piano, and then the sampler will let you play from its own or another MIDI keyboard as if you had all the notes from a piano. Of course, the pitch change algorithms used in these instruments introduce approximation errors when you sample just one note and then try to make an entire keyboard from it. Because of this problem, the more elaborate samplers are set up so that a series of notes can be sampled, so that the pitch change algorithm only needs to cover a few notes. In this case, the results can be extremely realistic.

All this is very exciting, and the music business today is undergoing rapid change as musicians learn to use the vast capabilities contained in MIDI systems. However, the interest to us is that these systems also embrace desktop computers. There are MIDI interface boards for PCs and a host of software that performs all kinds of functions. For example, the PC can function as a multitrack recorder, which records MIDI commands from many instruments in parallel. If these are played simultaneously by the computer, you can essentially play an entire orchestra from the PC. At the same time, the PC can edit the data to make changes or corrections, it can adjust the timing or the tuning, it can rearrange the measures — you name it. We have essentially a word processor for music. With samplers, the mass storage of the PC can hold samples, and the PC can function as a sample librarian.

As you may have figured out, MIDI is an alternate way to go if your interest in digital audio on a PC is for music. With a PC connected to a suitable bank of MIDI sound generators, you can produce high quality music of any style, and with far less data than any audio compression of real audio data could ever deliver. The underlying technologies used in the music industry are the same ones that we need for our PC audio system — sampling, processing, storage — and the rate of development for music can only help us get the audio hardware needed to make powerful and effective audio systems for PCs. In the future, we can expect that every personal computer will have built-in capability for realistic audio.

Summary

The points to remember about analog and digital audio are

- Realistic video is not effective without realistic audio.
- Audio is not easier than video — but it's different.
- An audio signal has no structure: this makes audio harder to compress than video.
- Audio is real time; there are no places where audio can be interrupted without it being heard.
- The audio industry is ahead of the video industry in terms of the use of digital techniques.
- To learn how to do something in digital audio, look at how they do it in the music industry.

5

Optical Storage Systems

We have already mentioned several of the industry trends that are bringing a computer-video system closer to reality. These are the advances in computer processing power and large-scale and very-large-scale digital integrated circuits. Yet another trend that is important to the use of video on computers is the growth of optical digital recording. This chapter will cover that subject.

The Compact Disc – Digital Audio (CD-DA) has made a remarkable penetration into the prerecorded home audio market. The 12 cm (4.72 in.) CD-DA disc is rapidly supplanting cassette tape and the LP record as the preferred prerecorded audio medium. Players have come down in price and are selling millions of units annually. All this success is very fortunate for the computer video business because the CD-DA has a spinoff which can be used for read-only storage of any kind of digital data. That spinoff is the CD-ROM, and it benefits from the technology development and cost reduction that was done initially for the audio business. Let's look at some of the numbers for the CD-DA to gain insight into the technology's use for general digital read-only storage.

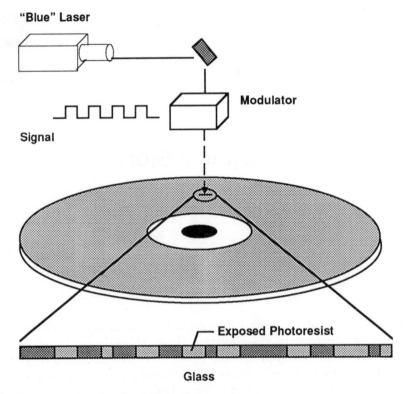

Recording

Recorded Master (Cross Section)

Figure 5.1   Optical mastering process (courtesy of 3M Optical
Recording)

The CD-DA plays stereo audio for up to 74 minutes in duration, and
it uses linear PCM at 44.1 kHz sampling rate with 16 bits per sample per
channel. Multiplying those numbers out shows that the CD-DA is
storing 764 megabytes of digitized audio data! All that on one side of a
disc only 12 cm (4.72 in.) in diameter that can be replicated by pressing
for a few dollars per copy. How does it do it? Laser optical recording
technology is the answer.

Laser optical recording has been around for more than ten years. It

Pit Dimensions

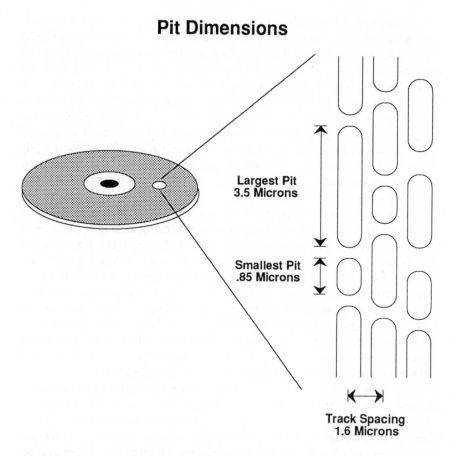

Largest Pit
3.5 Microns

Smallest Pit
.85 Microns

Track Spacing
1.6 Microns

Figure 5.2   **Compact disc pit and track dimensions (courtesy of 3M Optical Recording)**

is used in many industrial and military applications, but by far the greatest volume of use (before CD-DA) has been in the LaserVision™ video disc (LV).   LV is a read-only analog system which reproduces motion video and audio on a 12 in. disc. Since it uses a disc format, LV has the ability to quickly access any segment of its recorded audio/video material just like we do with computer disc drives. Fast random-access operation is required to create audio/video systems which are interactive.   The LV disc has built a market for itself in educational, point-of-sale, and certain specialized industrial markets.

The process of recording for read-only optical discs is called *master-ing*, and both LV and the Compact Disc use the same underlying technology for mastering. Figure 5.1 shows the basic approach. To make a master recording, a very small spot from a high-power laser is focused onto a rotating platter coated with photo-resist layer that is sensitive to the laser light. In LV recording, the analog video information is modulated onto an FM carrier in much the same way as for video magnetic recording. This FM-video is used to turn the laser on and off while the platter is rotating. After development of the photo-resist layer the result is a track of *pits* on the rotating platter where the laser exposed the photo-resist and *lands* where the photo-resist was not exposed to the laser light beam.

In order to make this track of pits and lands into a continuous medium, the laser must also be moved by mechanical means in a radial direction across the rotating platter, producing a spiral track starting at the inside of the platter and moving toward the outside. Figure 5.2 shows the pattern of pits, lands, and tracks. Because the circumference of the track increases as the laser moves toward the outside of the disc, it is necessary to reduce the rotation speed of the platter in order to keep the laser scanning speed constant. Operating the LV system this way can record an hour of video on a side. This mode of operation is called *constant linear velocity* (CLV) because the linear velocity of the track is held constant by changing the rotating speed of the platter.

Once a master platter has been made with the high-power laser, the master is replicated by metallic plating or other process to produce a *stamper*. The stamper is used in disc pressing machines to make low cost plastic copies of the original recording. The embossed surface of pits and lands is accurately reproduced on every copy. The replicated plastic copy is then plated with a reflective surface, and a clear protective layer is added to complete the process.

A player for optical discs uses an inexpensive low-power laser focused to a spot on the disc in a fashion similar to that in the mastering process. However, the playback optics is designed to reflect the playback laser's light off the disc onto a photodiode which converts the light into an electrical signal. Whenever the playback laser spot encounters a pit, there will be no reflection, whereas the lands will reflect the laser. Therefore, the output of the photodiode recovers the same pattern that was recorded. Additional optical and electronic features are necessary in the player to control the radial position of the laser spot to ensure that the playback will follow or *track* the recorded spiral of pits.

It is also possible to operate the LV system in a different mode, called *constant angular velocity* (CAV), where the rotation speed of the platter is synchronized to the television vertical frequency. This is advantageous for applications which desire to random-access the information on the disc, because the disc rotation can then be synchronized with the video so that all the vertical blanking intervals will be lined up at the same angular locations on the disc. In playback, particular frames may be selected by moving the laser scanning head in a radial direction, but vertical synchronization will always remain stable. However, to retain fidelity at the smallest track diameter, the CAV disc rotation speed must be made higher than CLV's average speed, and the result is that recording time is cut in half — 30 minutes for LV-CAV.

The CD-DA uses essentially the same technology as LV-CLV to do digital recording. The digital bits in encoded form are used to switch the recording laser on and off. For example, each pit location on the platter surface could represent a digital 1 whenever there is a pit and a digital 0 if there is a land at the location. (It really isn't that simple, but you get the idea.)

CD-ROM

The CD-DA technology for recording and reproducing digitized audio can be used to handle general digital data, and that is the objective of the CD-ROM standard. However, general data use implies some major differences in how the disc should be formatted and in the degree of data integrity that it provides. CD-DA is designed for use with audio segments that play for long times (up to the entire disc in length, sometimes) and only limited access to intermediate points in the information. This is referred to as *linear-play* operation and because of it, the CD-DA designers chose CLV mode. On the other hand, data use requires random-access to potentially any point on the disc, and the data may be in blocks of any size. The use of CLV in the CD-ROM causes it to be slow for random-accesses.

CD-DA operates successfully with digital data error rates of the order of 10^{-8} (one bit error in 100 million) coming from the disc. Error concealment strategies are included in the processing in the CD-DA player to make this degree of error rate not noticeable to the listener. However, that kind of error rate is not acceptable for general data use, and therefore additional error detection and correction (EDAC) techniques are included in CD-ROM to bring the error rate down to 10^{-12}. This rate is equivalent to only one bit error in every 200 discs!

In order to provide for full random-access and for EDAC, the CD-ROM arranges its data into *sectors* of 2352 bytes each. The sectors are recorded in a continuous string on the disc, with total capacity up to 330,000 sectors. (The total capacity of CD-ROMs is not completely standardized yet. Some drives quote capacity as low as 270,000 sectors, which corresponds to CD-DA playing time of only one hour.)

There are two modes for use of the sector data – Mode 1 provides 2048 bytes of user data per sector, and the other 304 bytes are devoted to bit synchronization, sector identification, error detection code, and error correction code. This is the mode usually used for data, and it provides the 10^{-12} error rate performance. Mode 2 eliminates the EDAC codes, giving 2328 bytes of user data space per sector, but at the raw error rate of the disc itself. Mode 2 is only usable for data (like PCM audio) which may tolerate the poorer error performance.

The 330,000 sectors of 2048 user data bytes in Mode 1 gives CD-ROM a maximum data storage capacity of 660 megabytes. (Mode 2 storage capacity is 750 megabytes.) Because of the CD-DA heritage of CD-ROM, where the consumer would specify play time in minutes and seconds, the sectors are not numbered in a simple series from 0 to 329,999, rather they are specified by minute, second, and sector number. There are 75 sectors per second at the CLV speed and the maximum playing time of different drives ranges from 60 minutes to 74 minutes. We will talk shortly about how an appropriate operating system will hide all of this numbering complexity from the user.

Having 75 sectors per second with 2048 bytes per sector means that the data rate of a CD-ROM in Mode 1 is 150 kbytes per second. This is faster than a floppy disc but slower than the hard discs used in PCs. Note that since the CD-ROM always operates at the same CLV speed, the data rate is a constant — it is not possible to get the data out any faster. In fact, it's not easy to make the data come out slower either — one must do repeated seeks to slow the data down. This constant data rate often leads to the need for buffering of CD-ROM data in RAM or in dedicated hardware in order to match the data rate to the requirements of the application.

Probably the greatest drawback of CD-ROM as a digital storage device is in the area of *access time*, the time required to move to a specified sector and begin reading data. CD-ROM access is slow compared to either floppy or hard disc, and the impact of access time on the operation of a CD-ROM application will almost always be an issue in the application design. To quantify access time, let's look at what is involved to access a specific sector. There are three parts to CD-ROM access time: *seek* time, *synchronization* time, and *rotational delay.*

Seek time is the time to physically move the laser head to the correct radius and achieve the correct rotational speed for that radius — it can take up to 1 second with some drives for a full seek from one end of the data to the other. Seek times are of course shorter if the radial distance to be moved is shorter.

Synchronization time is the time needed for the laser reading head to settle and lock in on the correct section of the spiral track. This time is short (milliseconds) unless the seek did not bring the head to exactly the correct track to begin with — in that case additional short distance seeking must be done to move to the correct track.

Rotational delay is the time for the disc to rotate until the desired sector comes under the reading head. CD-ROM rotates at 530 rpm at inner radius and slows down to 200 rpm at the outer radius. Therefore, the worst case of rotational delay is at the outside, where this delay could range up to 300 milliseconds.

Putting all of the above together, one can see that CD-ROM access time is a variable, depending on where you are on the disc and where you want to go, the drive performance for seek and synchronization, and how lucky you are with regard to rotational delay. Short distance accesses should be in the hundreds of milliseconds, but long accesses could go up to several seconds. This is clearly a factor which must be addressed in the design of application structure, data structures, data buffering, and CD-ROM disc layout. As in any good system design, we need to optimize the use of the system's resources, taking into account their strengths and weaknesses.

CD-ROM standards

The CD-ROM format as described above is laid down in a document called the *Yellow Book* published by Philips and Sony, the designers of the Compact Disc system. (The CD-DA standard is the *Red Book*.) The Yellow Book describes the physical format of the disc, and it allows data to be located by means of the sector addressing scheme. However, in any application of this format, there must be a higher-level specification about how any particular kind of data is organized into sectors and how a program will find those sectors. This is called the *logical format* for the disc. The Yellow Book alone would allow every user of the CD-ROM to create his or her own logical format, which would lead to discs being totally incompatible, even among different software companies' programs which are intended to perform similar functions. To propose a logical format standard, a group of manufacturers and software companies formed a committee called the *High Sierra Group* (named for the location of their first meeting — High Sierra Lodge, Lake Tahoe, California), which created a document that was submitted to the ISO for worldwide standardization. The ISO has completed their review, and the High Sierra proposal, with a few revisions, is now the ISO 9660 standard.

ISO 9660 is actually called a *file standard*, because it organizes data into files the same way data is normally organized on floppy and hard discs. This file standard is intended to be used with present operating systems, which will add their own CD-ROM driver software so that CD-ROM files can be accessed by applications and users exactly the same way they now access floppy and hard disc files. The same file standard can be used by different computers, such as Macintosh, IBM PC or compatibles and others. In the case of the MS-DOS operating system for personal computers, Microsoft has announced the Microsoft CD-ROM Extension (MSCDEX) software to extend MS-DOS to CD-ROM access. This software, which is being marketed by bundling it with CD-ROM drives and applications, makes the CD-ROM drive look like any other DOS disk device, except of course you cannot write to it.

ISO 9660 specifies certain data structures to be placed on a CD-ROM disc in addition to the data files themselves. These structures allow the operating system to find files by name. However, there are some special considerations that come into play when doing this for CD-ROM. First, in most operating systems for floppy or hard disks, the system asks the drive to perform several seeks before a file is retrieved. It does this in order to search a *directory tree* to find where the file is physically located on the disk. (Directories are themselves files, which are stored separately from the data files.) Directories contain all the information needed to associate file names with physical disk locations. On a computer that has a hierarchical directory system (subdirectories), there will be a directory file for each subdirectory on the disk. To find a file that is in a subdirectory several levels down in the directory tree, the system will have to seek to and read each subdirectory beginning with the root directory and going down to the one containing the desired file. On a CD-ROM, this number of seeks would take much too long. Therefore, the High Sierra Group proposed a scheme which allows a *path table* to be read into RAM when a CD-ROM disc is mounted; then, by reading the RAM path table to find a file's subdirectory location before seeking for the file, a single seek will move you directly to the subdirectory that contains the desired file. A second seek will be needed to begin reading the file itself.

The idea of the path table is possible because the CD-ROM is read-only. The subdirectory tree is determined at the time of mastering, and it will never change thereafter. For that reason, it is also possible to read the contents of the desired subdirectory into RAM (if you have enough RAM available), and subsequent reading of files in the same subdirectory can then be done with only one physical seek by the drive.

Having the logical disc format allows all applications to access their files in the same way; it goes a long way toward standardization of CD-ROM discs. But it does not say anything about what is inside of the files. Specifying the files themselves is yet another step to achieve complete standardization of CD-ROM discs. At this point we also come face to face with the question of whether file contents standardization requires one also to standardize the *system* which uses the contents of the files. There are several approaches to that question. One system which uses definition at the system level is *Compact Disc-Interactive* (CD-I) — we will discuss that next.

Compact Disc-Interactive (CD-I)

Overall CD-I concept

Just like CD-DA and CD-ROM, CD-I was created by Philips and Sony, and it is described by the *Green Book* document. The Green Book describes a complete system of disc format, player hardware, and system software for a low-cost audio/video/computer product, suitable for the consumer market as well as other markets where cost is a major issue. CD-I's disc format is compatible with the CD-ROM High Sierra scheme, but it goes much further to completely specify how the contents of the files are structured and what the characteristics of the associated hardware and software must be. Philips and Sony have licensed many companies to produce hardware and software for CD-I.

We will not be going into all of the details of CD-I in this book, because CD-I is not intended to be a personal computer. CD-I could be looked at as a specification for a new and different personal computer, but its developers have their eyes on the consumer market where it may be a good marketing approach to not emphasize the fact that a product contains a computer. In order to not look too much like a computer to the consumer market, fundamental computer features such as alpha-numeric keyboards and writeable mass storage are *options* on CD-I. But because CD-I is an important example of a CD-based interactive audio/video/computer system, you should understand its concepts, so we will cover it here at that level.

CD-I hardware

The fundamental CD-I hardware unit is called a *decoder*. In the simplest configuration (called the *base case* decoder) it is a single package containing the following units:

- CD disc drive, including CD-DA decoder
- CD control unit with sector decoder
- System board including:
  68000-family CPU
  RAM
  ROM containing CD-RTOS software
  Nonvolatile RAM
  Video custom VLSI chip
  Audio custom VLSI chip
  Miscellaneous glue logic
- User-interface pointing device
- Interface to RGB monitor or TV
- Power supply

This could be built into a unit approximately the size of a consumer VCR, which can be placed on top of a TV like we now do with VCRs. The CD-I decoder receives its programs and data on CD discs. CD-I applications must be packaged exclusively on CD discs, because there is no other kind of mass storage in the system. A CD-I base case decoder also is capable of playing CD-DA standard discs for audio only.

The CD-I CPU is a 68000-family processor running CD-I's unique operating system CD-RTOS (Compact Disc-Real Time Operating System). CD-RTOS is an outgrowth of the OS-9 operating system, which has been available for 68000 systems for some time. CD-RTOS is designed to have the real-time multitasking features required in audio and video systems. Including the details of the CPU and operating system in the specification is necessary because the program code for applications

Table 5.1 — CD-I Audio Modes

Mode	Sampling Rate per second	Bandwidth	Data Rate bytes/sec/chan
CD-DA stereo	44,100	20 kHz	171,000
ADPCM A	37,800	17 kHz	42,500
ADPCM B	37,800	17 kHz	21,300
ADPCM C	18,900	8.5 kHz	10,600

must be on the CD read-only disc and the program is written in (or compiled into) the CPU's native code. The hardware system is further specified to include unique audio and video features, which are implemented in custom VLSI chips. These chips will be used in all CD-I decoders, and therefore their detailed characteristics are written into the CD-I document.

The CD-I design provides for a full range of expansion possibilities or peripherals to be added to the base case decoder, which can provide a full PC capability (although with its own unique software environment). These include RAM expansion, various coprocessors, merging of external audio or video, floppy or hard disk storage, MIDI, printers and different video subsystems.

CD-I audio
The CD-I audio hardware provides for four different audio modes, which can be combined as necessary on the same application disc. The audio modes are tabulated in Table 5.1. The highest level is standard CD-DA audio, which takes over the entire CD data stream when it is being used. Then there are three lesser-quality modes based on ADPCM; a hardware ADPCM decoder is built into the base case system. With the lesser quality modes, the full CD data rate is not used, so there is opportunity to either have video and graphics with the audio or to have more than one channel of audio available.

CD-I video
The CD-I video hardware provides a wide range of video or graphics modes, also built into custom VLSI chips. Table 5.2 summarizes the CD-I video modes.

Table 5.2 — CD-I Video Summary (NTSC)

Mode	Resolution	bpp	Colors	Image Data
RGB 5:5:5	360x240	16	32,768	172,800
CLUT 8	360x240	8	256	86,400
CLUT 4	720x240	4	16	86,400
Run-Length	360x240	n.a.	128	10,000– 20,000
DYUV	360x240	18	262,144	86,400

The CD-I video hardware provides for two separate image planes in RAM, each of which can store a complete image in most of the above modes. The two planes can be combined in a keying or overlay fashion. In addition, there is a separate small image plane for a cursor image and a background plane which can display an incoming real-time image along with information from the other planes. All the image combining is done in hardware; it includes a variety of software-selectable modes.

The information in Table 5.2 is for still images. CD-I has capability for partial-screen or low frame rate motion video; however, the motion video performance is constrained by the data rate from CD and the lack of video hardware capable of sophisticated decompression of highly compressed real images. The Run-Length (RL) mode is supported in hardware and provides cartoonlike images which are highly compressed. Because of the low data per frame, full-screen RL images can be reproduced at frame rates in the range of 10 frames per second and thus they offer a degree of animation for 128-color images. It can be expected that competitive pressures will cause CD-I's limitations on motion video to be addressed in the future by more options on top of the base case system.

The DYUV mode is a simple compressed video format using a combination of DPCM and chrominance subsampling. It provides 24-bit color performance while using an average of only 8 bits per pixel.

To summarize then, CD-I is a system specification for an interactive audio, video, and computer system based on CD as the storage medium. CD-I has a wide range of capabilities and it is addressed to the consumer (home) market and other markets where low cost is important. However, many of those capabilities are cast in hardware of custom VLSI chips, and future enhancements inherently will require hardware changes or additions to the CD-I system.

Writeable Optical Media

The compact disc as discussed above is a read-only medium, but in the personal computer environment we also must have writeable or temporary storage — this requires another medium. Typical PC-based audio/ video systems will still use magnetic floppy and hard disks in addition to the CD-ROM. A system could be further simplified if we had a writeable optical medium which could store the vast data needed for audio and video and also could be used for local or temporary storage. That kind of product is going to happen. Technology now under development or in limited marketing promises a single optical medium in a few years which will fill all the mass storage needs of a PC.

One medium for writeable optical storage is the WORM (Write Once, Read Many) system. WORM drives for installation in PCs or as external peripherals are available on the market from several manufacturers in the price range from $2500 to $5000. WORM systems can write new data on their platters, but they cannot erase it, so every piece of new or modified data must be written to a new area of the platter. However, they are still useful, particularly as a medium to assemble the large quantities of data eventually intended to be placed on CD-ROM. The data capacity of a 5 1/4 in. worm drive is about 200 megabytes. So far there has not been standardization between different manufacturers — there are differences in formats and in the actual technology used. This market is in an early stage, and it should be approached with suitable caution.

Eraseable, Rewriteable Media

There are also some eraseable optical systems coming on the market. These are in an even earlier stage of development than the WORM; however, they are pointing to what is going to be available in more mature form in a few years. Because an eraseable system is more flexible than the WORM approach, rapid introduction of eraseable products could limit the application of WORM systems — system designers may choose to go with the more flexible approach. Of course this will also depend on the price tags associated with the different approaches.

Summary

A summary of the points made in this chapter on optical storage systems is

- Optical recording technology has been around for more than 10 years in the LaserVision analog video disc.
- Because of the success of the CD-DA audio system, low cost and reliable digital optical drives are now available.
- CD-ROM is the standard disc format of the CD for digital data.
- CD-ROM data rates are slower than hard discs but faster than floppy discs.
- CD-ROM access times are slower than both floppy discs and hard discs.
- The developers of the Compact Disc are continuing to expand the technology — CD-I is their latest version, an audio/video/computer system designed for cost-sensitive markets.
- Eraseable and rerecordable optical technology is under development and will be marketed in the next few years.

6

The PC and Digital Audio/Video Together

We have now covered separately all the elements needed for the combination of audio, video, and personal computers to create a multimedia display and processing system on our desktops. These are

Video technology — analog and digital
Audio technology — analog and digital
Large capacity optical storage systems

In addition, we have indicated several industry trends that will also contribute to the making of a desktop audio/video system:

Growth of PC processing power
Wider use of digital very-large-scale integrated circuits

In this chapter we will discuss the issues that arise when we bring all of these diverse technologies together to produce our desired system. We will also develop a conceptual specification for such a system.

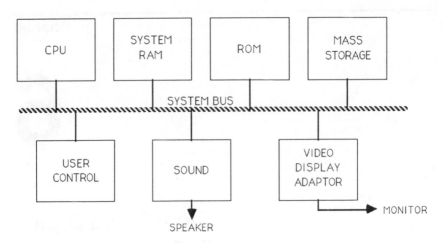

Figure 6.1   Typical personal computer architecture

PC System Architecture

Figure 6.1 shows a simplified block diagram for a PC/AT or compatible computer system. The heart of the system is the *system bus* to which all units are connected. The system bus delivers 16-bit data and 24-bit address to all units. It also provides other signals for memory control, interrupt control, and system clock. Since everything is in parallel on the system bus, any unit can talk to any other unit; but only one path can be active at any one time, under control of the CPU by means of address selection and bus control signals. In an 8 MHz PC/AT, the system bus operates at a maximum speed of 2.67 million cycles per second for 16-bit transfers. Because the CPU is involved in many kinds of bus transfers, there are not many operations that can actually run anywhere near maximum bus speed. For example, a block move in RAM using the fastest possible CPU instructions will run at about 800,000 cycles/second.

The video display requires a lot of memory activity to move data out for refresh of the screen. For example, if we have a VGA display at 640x480 resolution, 4 bpp, refreshed at 60/second, we are talking about accessing 9.22 million bytes/second. The system bus could not handle this rate if it did nothing else. Therefore, video display adaptors include their own memory for storing the screen bitmap, and they have separate circuits for refreshing the monitor from the bitmap memory without making use of the system bus. The screen bitmap memory is memory-mapped into the CPU's address space (meaning that the hardware is set up to make the screen memory appear in the same address space as the system RAM) so that the CPU can write to it to enter information onto the display. That operation, of course, does make use of the system bus.

Mass storage passes all of its data through the system bus to load into system RAM or the screen display RAM on the video display adaptor. The speed of mass storage devices is determined by the rate at which they can read from their storage medium. This operation also typically requires a lot of CPU involvement, and net data rates are much lower than system bus speeds. For example, hard disk maximum data rates are around 500,000 bytes/second while a contiguous block of data is being read from the media (so that there are no seeks involved). When the actions of the operating system are included, it becomes difficult to maintain an average hard disk data rate of more than about 150,000 bytes/second for any length of time. Further, because a single-user operating system has nothing else to do while it is waiting for disc I/O, it is common for single-user operating systems such as MS-DOS to simply tie up all the bus cycles during any mass storage access. This is an area that can be a major problem in a real-time audio or video system. We'll come back to this issue.

Data Rate Considerations

In the previous chapters we established that realistic video images for display in a desktop situation will require a pixel resolution of 512x480 or higher and a bpp value of at least 16 bits. This means that a single full-screen image takes 491,520 bytes. At the hard disk average data rate quoted above, it will take about 3.3 seconds to retrieve that size image from hard disk, even if we write from hard disk directly into the screen memory of a realistic video display adaptor. Therefore, we will sit for 3.3 seconds watching a single image being written onto the screen. The other way of saying that is that we can provide motion video at a frame rate of 1/3 frame per second — which no one is going to recognize as motion. There are several issues that this example brings up.

Obviously, if there were less data needed for a frame of video, everything would be easier. For example, if we reduce the resolution to 256x240, which would make a soft picture in the desktop viewing situation, but would be all right for television-style viewing, the data per frame is reduced by a factor of 4. So in that situation, we could move up to 4/3 frames per second for motion — still not too great. But suppose we also had a way of compressing the data by a factor of 10, and assume that the decompression process didn't take any time (which is a challenge in itself), then we could display the 256x240 images at 13 frames per second — which is beginning to look like smooth motion. So one direction to go is to use compressed images and include a very fast decompression capability in our system. We will discuss a system in the next chapter which can achieve even more than a 10:1 compression ratio.

Another way to reduce the data in order to have better motion rendition is to show motion in a partial screen or a window configuration. When we are dealing with a fixed data rate like the 150,000 bytes/second rate used in the above examples, we can relate that data rate to pixels/second. If we are uncompressed at 16 bpp, then the 150,000 bytes/second data rate represents 75,000 pixels/second. If we have an image window which contains 75,000 pixels (say 316x237), then we can display 1 frame per second in that window at the 75,000 pixel/second rate. If we want to display at 30 frames/second using the same pixel rate, then our window size must be reduced to 58x43, which contains only 2500 pixels. So another approach is to reduce the size of a motion window until we can handle the pixel rate. Many applications can use this approach because there usually is a need to have other things on the screen at the same time anyway. Making use of this kind of trade-off calls for the system design to allow great flexibility in the way images can be intermixed and in the resolution numbers for displaying them.

Combining the ideas of video compression and motion in windows is also a good possibility. With compression, we can effectively reduce the average bpp value in the compressed data, so the same amount of data can represent more pixels. In the example used earlier, where we said we might compress by a factor of 10, we would get an average bpp value of 1.6, so now our 150,000 bytes/second data rate from the hard disk could pass 750,000 pixels/second. At that pixel rate, we could do 30 frame/second motion video in a window of 182x137. All of these examples are simply making the point that our realistic video system design should allow the flexibility to combine methods to achieve motion video in any size of window and with any kind of compression.

The same kind of discussion that we have gone through above can be applied to audio. Of course, the discrepancies are not as large between the required data rates and what the system capability is. For example, the 150,000 byte/second rate used above will deliver CD-DA quality digital audio, which is the best quality there is. However, the problem comes about when we realize that the same 150,000 bytes/second must carry both audio *and* video. And since the video is so demanding of data rate, we should allocate as much data rate as possible to video. Thus we want to squeeze the audio data as hard as possible just to achieve the maximum data rate for video. This means that the system should also include capability for handling compressed audio, and the audio capability needs to have great flexibility so we can tailor the use of the data rate resource dynamically within an application, depending on the immediate needs for audio and video.

Processing Power for Video

In the foregoing discussion, we spoke only of moving video and audio data from mass storage into memory. We really would like to do much more than that in order to exploit the capability of a digital system to manipulate the video and, to a lesser degree, the audio. However, manipulation of data by a computer involves some kind of processor performing operations on the data before it is displayed. Since all data is passing through the system bus under control of the CPU, it would be simple to add to the software running on the CPU to perform any operations on the data we would want. However, for several reasons, that turns out to not be a very good idea. Consider the requirements for manipulation of video, for example.

Let's say we simply want to change the size of an image: we have a 512x480 image on disk, which we want to display in a quarter-screen window on a 512x480 screen. That means we need to drop every other pixel and every other line of pixels as we move the image into the screen display memory. There are several ways to set up this operation with the system CPU, but to make it as simple as possible, assume that we have already retrieved the original image from disk and it is in RAM somewhere. The shrink operation will then take place as a separate task to be done while we move the image from that storage memory area to the screen display area of memory.

(We're going to resort to some 80286 assembly language here, but don't worry, you'll get the point even if you don't know assembly language. The reason for using assembly language rather than some possibly more familiar high level language is that assembly language lets you use the 80286 in the most efficient way possible.) In the 80286 CPU, the 2:1 image shrink operation can be done with the MOVS 80286 instruction, which can move data (in either byte or word format) from one region of memory to another region. We can arrange two loops where the inner loop moves pixels within a line and the outer loop increments the source address to skip every other line. MOVS uses two registers to index the addresses of these two locations by the appropriate amounts to move sequential bytes or words. However, since we are going to skip every other pixel in the source image, we must add another instruction to the inner loop to increment the source address by another pixel. The inner loop (where most of the execution time will be) must then include three instructions — MOVS to move the data and increment, an extra INC to skip pixels, and a LOOP conditional jump to close the loop until the count of pixels reaches zero. The inner loop will look like:

```
; (SET UP SOURCE AND DEST. ADDRESSES FOR MOVS)
; (LOAD PIXEL COUNT FOR LOOP INSTRUCTION)
INNER:        MOVS            ; 5 clocks
              INC    SOURCE ; 2 clocks
              LOOP   INNER  ; 8 clocks
; (PROCEED WITH OUTER LOOP TO SKIP A LINE)
```

By adding up the execution times of the three instructions above, we find that each pass through the inner loop will take 15 CPU clock cycles. With an 8 MHz clock, that is 1.87 microseconds per pass. The inner loop will be traversed once for each pixel of the output image, so since the output image is 256x240, the inner loop takes a total of 115 milliseconds. Adding maybe 10% for initialization and the outer loop's cycle time, this image shrink operation using the 80286 CPU will take about 0.13 seconds, or about 7 frames/second. That doesn't sound too bad, but this is an extremely simple operation, done on a relatively low resolution image, and using only three CPU instructions per pixel. If we get into more complex operations, such as a shrink that is not exactly 2:1, which would require some pixel averaging or interpolation, we can easily require tens of instructions per pixel. To do such operations on high resolution images with the 80286 can take tens of seconds. And remember, when the 80286 is running in tight loops doing pixel operations, it is not doing anything else, so everything else in the application has to stop.

A solution to the limitations of using the system CPU for video operations is to include a *video coprocessor,* which will be independent of the system CPU and will operate in parallel to perform all video manipulation. Chips are on the market that can do that, and they greatly increase the capability of the system at relatively low cost. Once we have decided to accept the cost of a video coprocessor, however, there are a lot of other things you could ask it to do. It could manage the screen display memory for us and handle screen refresh and prioritization of memory accesses by the system CPU. It could also have capability to do many of the algorithms we would like to apply to images. In that sense, it would really be great if the video coprocessor had its own instruction set — tailored for video use — and could be *programmed* for anything we want it to do. Now we really have something that can survive the tremendous appetite for innovation in the PC industry — new video ideas would all be in *software,* and the hardware could be designed to have a long life in the marketplace. That is exactly the same thing that has kept the PC itself going for so long — it takes years of software innovation to bring out all the latent capabilities hiding in well-designed hardware. A good hardware platform that can respond to all the things that software designers will think of once they get going with the system — that is the basis for a successful PC product. At the same time, the hardware itself can improve over time while maintaining compatibility with all software, as the PC has gone from the 8088 to the 80386 CPU, this is exactly what has happened. It applies to video peripherals just as much as it applies to the PC as a whole.

The idea of a powerful and programmable video processor is the key to the successful marriage of video with the personal computer. Although it may well be easier (from the hardware design standpoint) to cast a lot of the video algorithms in special-purpose hardware — still using VLSI chips — one can be sure that the limitations of such an approach will be quickly reached by the industry. The digital video field is still young, and the best algorithmic approaches have probably not even been invented for many of the important tasks in digital video manipulation. It is far too soon to think that any group of hardware designers can make all the right choices and put together the perfect combination of features that will satisfy the market for years to come. We must make the world-beating video system fully programmable in software!

We should divert here for a moment and use the preceding example to define some more terms. In the example the 80286 was executing three instructions in 1.87 microseconds. That means that the example was running at 3/1.87= 1.6 *million instructions per second* (MIPS). The MIPS rating for a CPU is one measure of speed. However, it should be pointed out that the particular 80286 instructions used in this example are simple ones which do not use many clock cycles per instruction — it would not be possible to reach 1.6 MIPS in a generalized 80286 program that exercised the full instruction set of that CPU. Another parameter that is of interest is that we were doing one *pixel operation* in 1.87 microseconds, the pixel operation was a 2:1 shrink. Therefore the pixel speed rating was 1/1.87=0.53 million pixel operations per second (MPOPS). And it can be looked at one other way, which is the ratio of MIPS to MPOPS, which in this case is three. The example is a very simple pixel operation, but the 80286 CPU does not have a particularly suitable instruction set, which gives a high value of MIPS/MPOPS. A video coprocessor could very well have special instructions for doing the kinds of tasks needed in pixel operations, and such a CPU would have lower MIPS/MPOPS ratios than the 80286.

The Needs of Digital Audio

All of the above discussion of digital video pertain equally to digital audio. The same arguments for a programmable coprocessor that we expressed for video apply just as much for an audio coprocessor. We will not repeat all of that again, but remember that the audio resource is nearly as important as the video one for any system that is going to create realistic pictures. Also, the arguments for innovation through software apply equally well to the audio system.

Mass Storage Needs of Digital Video

We have talked so far only about the use of hard disk for storing digital video. That is an excellent approach, but the limits are quickly reached. Today a 40-megabyte (meg) hard disk is pretty common on PC/AT class of systems. In terms of the numbers we have been expressing above, the 40-meg disk stores 83 512x480 16-bpp still images. If we do motion video using any of the sets of numbers discussed above (which are based on the 150,000 byte/second data rate), 40 megs will play for 4.5 minutes if we use the entire disk for video. Thus you can see that 40 megabytes doesn't go very far in the video business, even with data compression. The audio situation is similar — if we use 16,000 byte/second audio, the 40-meg hard disk will store only 42 minutes of audio. We need much more storage capacity!

Of course, we can go to larger hard disks — they are available up into the hundreds of megabytes for use with PC/AT class of machines. This will work for many applications, but it gets expensive, because prices go up pretty fast. The real answer is in optical storage technology — CD-ROM, WORM, or the erasable optical storage devices that are coming soon. CD-ROM, particularly, is low in cost — the drives are very similar to the low cost CD-DA audio players, and they should soon cost about the same as the 40-meg hard disks. Furthermore, the data capacity goes over 600 megabytes per disk and the disks are removable, so we get even larger capacity by changing disks. Of course, there always has to be a catch — CD-ROM is, as the name says, *read-only*. To a computer person, this would seem to be a killer — and for some applications, it is. However, for video and audio, there are many applications where read-only is perfectly fine.

Writing of audio or video to storage is needed only if the user is going to produce his or her own new material. The creation of new audio or video implies that the user's system contains digitizers and that the user has the appropriate cameras, VCRs, and audio equipment needed to originate audio or video. There are many kinds of application that have a need to display audio and video from prerecorded sources but have no need for anything new to be created. These applications will use audio and video from read-only media as a resource to be called up as required to support whatever the user is doing. They may also use a hard drive to store non-A/V material that the user creates. CD-ROM is an ideal medium in such cases.

The existence of programmable video and audio processors also expands the scope of what can be done with read-only storage because the material from storage can be manipulated before it is presented to the user. Application designers who carefully think through the use of manipulation for prerecorded material in real time can greatly magnify the range of different effects which can be produced from what is held in the read-only storage. An extreme case of this is the interior design application described in Chapter 1, which uses the synthetic video technique to create realistic renderings of room designs. For that application, a CD-ROM holds only the *pieces* of an image, in the form of furniture 3-D models and separate video textures for all the coverings.

The application in real time assembles the pieces to generate realistic images, chosen from an infinity of possible room designs.

Applications that do have to store audio or video locally can still do that (with the appropriate origination equipment) on hard disk with only the limitation of data capacity to deal with.

RAM Needs of Digital Video

Storing digital video images for display takes a lot of RAM. Our example image of 512x480 pixels at 16 bpp uses 480 kilobytes of RAM to store one image. That is already too large to put in the address space of the PC/ AT running under MS-DOS, which is 640 kbytes for everything. In fact, most applications running with resolutions in the range quoted above will need even more RAM for video functions. In order to get around the slow loading of an image from disk (3.3 seconds for the above example), we could very well wish to load the image ahead of time to an area of RAM that is not being displayed. Then to bring the image to the screen quickly when the application calls for it, we could either do a move in memory to copy the image into the part of RAM being displayed or we could tell the display hardware to change its base address to the location of the off-screen image. Either way, we are going to need 960 kilobytes of RAM to do that. In this day and age, 1 megabyte of RAM for storage of images in a video system has to be considered the minimum value, and many applications will want more than that.

The large size of video memory space has several other implications. First, we need to figure out how the CPU can address such a large space, as long as we are running under MS-DOS with its 640-kbyte limitation. The usual solution is to *bank switch* the video memory into part of the 640-kbyte address space. For example, we might allocate 64 kbytes of system address space to access the video RAM. A 1-megabyte video memory could then be accessed in 16 *segments* of 64 kbytes each, with the segment selection under the control of an I/O port. The system software could include some subroutines which allow the definition of a separate video memory address which goes up to 1 meg (or more), and whenever video memory is accessed, the system subroutines would calculate the segment number from the video memory address and tell the bank switching port to go to that segment. Anyone accessing video memory by calling that subroutine would not have to be concerned with the bank switching process — it would be invisible to the calling program. Of course, the bank switching process adds a little overhead to all accesses to video memory, but if it is used carefully, it is not a serious problem. This approach is provided in many PCs to expand memory by use of the Expanded Memory System (EMS).

Another implication of the need for megabytes of video memory is that any video coprocessor should be able to address directly (without any bank switching) a large amount of memory. Sixteen megabytes would seem to be a minimum value today.

There are three classes of video memory access needed: by the system CPU for loading of image data, by the video coprocessor for video manipulation, and by the display refresh circuits for feeding data to the display. The display refresh requirement is usually by far the greatest number of accesses. Taking again our 512x480 16-bpp image, if we refresh the display at 60 per second, we will need a data rate of 28.8 megabytes/second out of video memory for display refresh. It would be extremely difficult to achieve such a rate from conventional dynamic RAM (*DRAM*) chips, even without considering the needs of the other two classes of access required. Fortunately, memory chip designers have also seen this problem and a variation of the standard DRAM chip is now available that is *dual-ported*. These dual-ported RAM chips are intended for video display refresh, and they have a second output path which includes a serial shift register that can easily reach the high rates we need. Better still, the high-speed output is completely separate from the usual random-access control port of a standard DRAM chip, so the system CPU and video CPU can access through the standard port, and they will never have to wait for display refresh actions.

Since the dual-ported DRAMs are made specifically for video display purposes, they are called *video RAM* (VRAM). VRAM chips are currently available from several manufacturers in 256-kilobit and 1-megabit/chip sizes. Some versions of the VRAM design also allow the shift register to be used for high-speed input to the RAM array. VRAM is an important area of technology which will continue to develop; we can expect that the cost/megabyte of video memory will continue to drop for some time.

I/O Devices for Digital Video

There are no particularly unusual user interfaces required for a digital video system. Conventional devices such as keyboard, mouse, tablet, or joystick all have their place in digital video systems. One device that is not common on personal computers, but is common in video disc based interactive video systems, is the touch screen. This is a device that is built into the display monitor or may be an overlay over the monitor screen; it responds to the user's touch by returning the position coordinates where the user is touching. Typically the output of such devices would enter the system via a RS-232 serial port, and special driver software would translate the output of the touch screen into something that the application software can understand. Because of the use of this technology in interactive video systems, we can expect that it will also be used in PC-video systems.

Software for Digital Video/Audio

In this chapter we have been building up the specification for a system which will add audio and video coprocessors with their own memory and other support to a PC AT so that audio and video can operate in parallel with the PC's other functions. However, the PC is still responsible for movement of data from mass storage to the audio and video subsystems. All of this is clearly going to produce some major demands on the software of the PC AT. If audio and video are truly going to run in parallel with other PC functions, we are talking about what is called *multi-tasking* in the computer world. Furthermore, we have already established that there are things about audio and video which cannot be delayed by other things going on in the computer — this means that we actually want *real-time* multitasking.

A computer operating system can achieve real-time multitasking by use of the CPU's interrupt features. We mentioned CD-I's CD-RTOS in the previous chapter — it is a real-time multitasking system. However, in the PC environment, the operating system most widely used is MS-DOS, which does not support either real-time or multitasking. Of course, we could throw out MS-DOS and go to a different system which already had the features we need — there are some candidates out there. This would be a traumatic step, and it would have to pay the price that all the PC software now in use would not run. That price is too high — it would severely limit the growth of audio/video use on PCs.

A better approach is to find a way to include the real-time and multitasking features that we need for audio/video *inside* of audio/video applications that run under standard MS-DOS. Then all of the existing PC software could still run in the normal way on the system, and new applications with audio/video would bring in their new operating system extensions with them. This seems like the best of both worlds. Note, however, that it has one limitation which ultimately we might not like to have — it is not true multitasking in the sense that we cannot run standard PC applications and new audio/video applications *at the same time*. The multitasking exists only for things which happen inside audio/video applications, and because the real-time feature requires that the audio/video application take over the interrupt structure of the system, another application which needed interrupts could not run simultaneously. So in the long term, this approach is a compromise. Ultimately, we should have a new operating system written from the ground up to contain the features needed for audio/video while retaining compatibility with MS-DOS. That is a large task, which would severely hold back the growth of PC audio/video if we wait for it—but it is on the way in new systems such as Microsoft Windows and OS-2.

Summary

We have now completed the discussions that define the basic requirements for an audio/video system for the PC environment. Some general approaches have been established for the elements of such a system which we would hope could be implemented in custom VLSI at a price that would be consistent with other PC add-in features. The points we have made are

- Audio and video must both be *digital*.
- Large-capacity, low-cost mass storage will be necessary. CD-ROM is a good choice, but the system design should not be tied only to this, because other suitable storage devices are coming in the near future.
- Data *compression* is required for both audio and video. Video compression performance should support full-screen full-motion video operation at hard disk or CD-ROM data rates.
- Algorithms for audio/video compression and processing should be in *software*, so that improvements coming out of current and future algorithm development work will not obsolete the hardware.

- Software should use existing operating systems and *add* the features needed for audio and video.

The preceding list may look like a very ambitious set of objectives, but in the next chapter we will begin the description of a system which was designed for these objectives and is now available — DVI Technology.

7

Introduction to DVI Technology

The previous chapters have developed the need for a PC-based interactive all-digital multimedia system that could play television-style video and sound as well as do all the things with graphics and sound that personal computers now do. We also reviewed the technologies that are involved in such a system and concluded with a set of objectives for a system. Such a list of objectives was being thought through in 1984 at the David Sarnoff Research Center in Princeton, New Jersey. That laboratory, then the central research facility for RCA corporation, has a long history in the development of audio and video systems, including most of the fundamentals for the color television systems now used worldwide. The Digital Products Research Group at the Sarnoff Center had already amassed years of experience in interactive video, and they were about to embark on development of an all-digital interactive video system. That work involved approximately 50 very dedicated people, who carried the project through the transition of the GE acqusition of RCA, and several other ripples, to bring it to the point of demonstration of actual hardware and software in early 1987. The most remarkable feature of the system they developed was that it played full motion digital video from a standard CD-ROM disc.

The technology, named Digital Video Interactive (DVI) Technology, was first shown publicly at the Second Microsoft CD-ROM Conference in Seattle, Washington, on March 1, 1987. DVI was shown as a technology development, with market plans still to be completed. The first public announcement was a 40-minute presentation to an audience of 1200 people at the conference. The audiovisual material for this presentation was done entirely with the DVI system, displayed on a large screen projector, showing slides from hard disk, motion video and audio from CD-ROM, and application demonstrations. The audience responded with an ovation to a degree that was unheard of for a technical meeting. Throughout the remainder of the conference there were five DVI systems operating continuously in an exhibit booth and a demonstration suite. Subsequent industry and press reaction was overwhelming — DVI clearly contained the features in interactive audio/video that the industry was looking for.

In the months following the announcement, an internal venture, the GE DVI Technology venture was established in GE to manage the continuing development of DVI. (GE became the owner of DVI when it acquired RCA in 1986.) A number of key decisions were made by the technical team and GE management.

- DVI will be brought to market as a general technology for audio/video use on any type of computer.
- The first platform will be the IBM PC AT and compatible computers.
- A broad group of hardware manufacturers will be sought to develop and produce additional hardware systems.
- Technical work will continue at the Sarnoff Center to complete product design of a DVI board set for PC AT computers. This board set will be marketed for both software development and end use.
- The DVI chip set will be taken to a full-custom design for production.
- Work will continue on improvement of the compression algorithms.

- A comprehensive developer's software package will be completed to allow DVI application development on the PC AT platform.
- Customer support will be provided to assist software development by independent software vendors, addressed primarily at vertical markets.

Although GE was the owner of the technology, the laboratory where all the people work was given in April, 1987 to SRI International, an independent research corporation headquartered in Menlo Park, California. The technical work continued at the same location, which is now called the David Sarnoff Research Center — a wholly-owned subsidiary of SRI International, under contract with GE. But it was still the same group of people.

In October 1988, GE announced that it had sold DVI Technology to Intel Corporation. Intel is the developer and manufacturer of the CPU chips (8088, 80286, 80386, etc.) used in the IBM PC, XT, AT, PS/2 and compatible computers. Although Intel is headquartered in Santa Clara, CA, they set up the Intel Princeton Operation as a separate business unit with its own building at Plainsboro, N.J. Most of the original DVI personnel are now employees of Intel, and they are continuing to pursue the commercialization strategy described above. Intel has substantially expanded the DVI staff and is committed to supporting the DVI group to make DVI the interactive audio/video standard for personal computers.

This chapter will introduce you to DVI Technology, which is the world's first example of a technology designed to meet the objectives for a PC-based interactive audio/video system that we covered earlier in this book. DVI Technology will be used in subsequent chapters to describe how various interactive video capabilities are accomplished with the system. We will begin with discussion of some general features and concepts of DVI Technology.

What Is DVI Technology?

As was explained when it was first introduced to the industry, DVI Technology is not a system, it is a *technology*. That means that many different systems could be bulit using the technology.

DVI Technology actually consists of four unique elements:

- A custom VLSI chip set, which is the heart of the video system
- A specification for a runtime software interface
- Some audio/video data file formats   .
- Compression and decompression algorithms

Using these four elements, DVI Technology could be implemented on any computer system that has sufficient power and storage capability. Hardware must be designed for that system to interface the DVI chips and to provide other features called for by the runtime software specification. Then software must be written in the native language of the host system to implement the DVI runtime software interface. This is a large design task, but it needs to be done only once for each computer type that uses DVI technology. These steps have so far been completed for two systems or *platforms* — IBM PS/2 computers, and PC AT compatible computers. Those implementations will be described in this book. Intel is also working with several DVI hardware component customers who are undertaking implementation of DVI Technology on their own specialized computer platforms.

While DVI Technology can be used in many systems, the PS/2 and PC AT version of DVI which is described specifically here, is simply the first embodiment of the technology. The PS/2 or the PC AT is the platform, and the DVI hardware and software add the DVI features to that platform. This concept is extremely important because it affects the approach to standardization for DVI Technology — which will be done in such a way that many systems are possible within the standard. The purpose of that is simply to attain the widest possible use of DVI Technology, and thus the lowest costs for everyone. DVI standards will contain as little hardware specification as possible to leave room for different system platforms and for future improvement of the hardware.

The previous paragraphs should not be interpreted to mean that it is easy to port DVI Technology to a different platform. Implementing DVI Technology on a different system, for example, one that uses a 68000 CPU instead of the 80286 or 80386, will mean major software work (many programmer-years) to rewrite and recompile the DVI system software to the different instruction set and the different environment. The DVI standardizing approach allows this to be done as long as the new system design brings the DVI software interface on the new platform up to the same set of functions that are defined now for the PS/2 and the PC AT. The resulting benefit is that the software application source code written for the PS/2 and PC AT version of DVI Technology can be ported to a new platform mostly by recompiling. (The previous discussion applies only to the parts of an application that are related to DVI Technology. Because of differences between operating systems of widely different platforms, parts of an application that call operating system functions directly will have to be completely changed — recompiling will not be enough.) Hardware porting to a different platform is largely concerned with matching the generalized bus interface of the DVI chips to the bus structure of a different machine and is a significant, but not difficult, hardware design task.

DVI Runtime Software Interface

The DVI software provides for all DVI functions to be activated by making C language function calls to the runtime software. An application for DVI Technology is best written in the C language, because all of the capabilities of the technology will then be easily accessible. The DVI runtime software functions themselves are written in C for the most part, with assembly language used where necessary for better performance.

There are higher-level authoring systems in existence that were designed for making interactive video disc applications. These systems have potential for simplifying the design of DVI applications because they already include approaches to deal with audio and video information. However, the authoring systems for video disc do not include capabilities for many of the new features of DVI Technology such as synthetic video, video manipulation, and high-speed graphics. In spite of that, several authoring systems specifically designed for DVI Technology are now on the market and will be described in more detail in Chapter 15.

Because the language of choice for DVI programming is C, we will be using a lot of C later in this book. If you already know C, that is great; but if you don't, you will find Appendix B helpful. It is a thumbnail course in C — but it does assume you know something about programming in another language such as BASIC or Pascal. Appendix B will bring you up to a point where you will be able to read and understand what is going on in the C examples we will be presenting. The purpose of those examples is to teach the concepts of audio/video programming, specifically for DVI, but they are not always complete programs by themselves. Much more C experience than you can get from Appendix B is needed to write an actual program, as any programmer will know. However if you are not a programmer, the information here will let you communicate with the people who are doing your programming and let you understand what they are up against.

DVI Architecture — the Chip Set

The heart of DVI Technology is the video coprocessor, which is implemented with two VLSI chips. This chip set is called the i750™ Video/graphics processors. The two chips in the set are the 82750 PA, the *pixel processor*, and the 82750 DA, the *display processor*. They are shown in Figure 7.1. Together, these two chips contain over 265,000 transistors to accomplish their functions. They were originally designed with a new computer-aided design technology called *silicon compilation* — a technique which allows a chip designer to specify logic functions and interconnections, after which the silicon compiler software will work out the actual design of a chip to perform those functions. It's not really that simple, but the process does substantially speed up the design of extremely complex VLSI chips. Silicon compiler technology speeds up getting the first working chips, but when volume production is expected, the first chips are redone in a full-custom design to reach the smallest die size and the lowest cost in production. Intel is doing that for the i750 Chip Set.

The 82750 DA display processor
The i750 chips are designed to use VRAM for their image memory. The pixel processor uses the random-access port, and the display processor uses the serial port. The display processor has the main function of taking data from a bitmap in VRAM and converting the bitmap data into a digital output signal which feeds a triple D/A converter to drive a color RGB monitor. The display processor knows about bitmaps and about a variety of bpp formats ranging from 8 to 32 bpp, and it operates at various resolutions up to 1024x512. It can deliver output signals that will be compatible with either NTSC or PAL interlaced sync or in many other interlaced or noninterlaced sync formats.

Figure 7.1   **Photograph of the i750 DVI Video/Graphics processor chips**

The 82750 PA pixel processor

The pixel processor is a special microprocessor which runs its own kind of software for performing operations on digital images stored in RAM. It is designed to work with VRAM for the image storage, and it runs at a speed of 12.5 MIPS. Furthermore, the pixel processor has an instruction set which already knows about pixels and bitmaps, and thus it is able to perform complex pixel manipulations with fewer instructions/pixel than a more general processor like the 80286. There are a lot of things which contribute to reaching the high speed of the 82750 PA, but one very important factor is that the chip executes its programs from its own internal RAM, so there is no slowdown caused by accessing external RAM for the program.

Programs for the pixel processor are written in *microcode*, which is a special kind of machine language that uses a wide word coding format to allow each of the individual parts of the internal chip architecture to be separately controlled for each instruction of the program. All video algorithms that run on the pixel processor must be written in microcode, which is loaded into the RAM on the chip as it is needed. Microcode programming is an art in itself, and it's beyond the scope of this book. However, the DVI runtime software takes care of the handling of microcode so that most programmers do not even need to know it's there. There is already an extensive library of microcoded functions for the 82750 PA available in the DVI software, and the system is set up so that this can be (and is being) expanded as more functions are developed in microcode. (Many of the functions in the microcode library will be described later.)

Hardware Packaging

The DVI hardware for the PS/2 and PC AT compatible environments consists of a single board available in either PC AT bus configuration or Micro Channel bus configuration. These boards are called the ActionMedia™ 750 Delivery Boards, and they are intended for end-user applications of DVI Technology. There is an optional second board for audio and video capture. It is called the ActionMedia 750 Capture Board and it is also available in both bus configurations. Using both boards together allows a full development workstation for DVI Technology to be built. The DVI development software is available as the ActionMedia Development Software, and a complete development platform is known as the ActionMedia 750 Application Development Platform (ADP).

ActionMedia™ delivery board

The i750 chip set has been designed onto a single add-in board which fits in one slot of a PS/2 or PC AT-compatible system. Figure 7.2 shows a photograph of the delivery board for a PS/2. The delivery board includes either 1 or 2 megabytes of VRAM for use by the i750 chip set, an interface for CD-ROM, and a complete audio processing system with dual audio outputs. The delivery board is designed to drive an analog RGB color monitor for the DVI display. The display on this monitor is completely independent of the usual PC standard output display (MDA, CGA, or EGA), so a DVI system can have two monitors. However, you can also use it in a single-monitor configuration — with a VGA display, standard PC VGA screens and DVI screens can show on the same display, even at the same time.

Delivery board audio

The audio capability on the delivery board uses a Texas Instruments TMS320C10 digital signal processing chip for the audio coprocessor. That chip is a general-purpose digital signal processing device that operates at 8 MIPS, and it is sufficiently powerful to accomplish all the audio functions needed for DVI Technology. However, it is not even close to the 82750 PA in power for video, because of the lack of any instructions like the PA's special instructions for pixel processing. The audio

Figure 7.2   Photograph of ActionMedia 750 Delivery Board (bottom), and ActionMedia 750 Capture Board (top). Boards shown are for IBM PS/2.

system also contains two output channels with D/A converters and programmable filters.

CD-ROM interface
The delivery board also includes a port to connect CD-ROM drives. With this interface, it is posible to bring CD-ROM input directly into VRAM without using any of the host computer bus bandwidth at all.

ActionMedia™ capture board
The ActionMedia capture board for PS/2 platforms is also shown in Figure 7.2. This board provides video capture from RGB inputs, and two-channel audio capture from line level inputs. For capturing video from an NTSC input, an external NTSC decoder is required.

Standards

Any technology which is intended for eventual broad application in millions of PCs around the world must have viable standards. This is necessary so that users can have the assurance that their purchases will not be quickly obsoleted by arbitrary changes in new products, and so that multiple manufacturers can confidently enter the market in order to create the competitive environment which is so necessary for large scale market growth. In the PC market, there is also the matter of software compatibility between different platforms and manufacturers. Software developers must be assured that their products will work with all hardware, regardless of manufacturer. Good standards can solve all these problems.

There is no single standardizing body which can (or would) address all the standardizing issues raised by a technology like DVI Technology. The manufacturer of the technology (Intel in this case) must take the initiative in first documenting the technology so that standardizing groups can understand it, and then to present the technology to the appropriate bodies. However, this is not going to cover everything — certain aspects of the technology will remain the manufacturer's responsibility, and the manufacturer himself must publish and support the standard. This is the case with respect to DVI hardware — Intel is responsible to see that the hardware is properly documented and that non-compatible changes do not happen. This responsibility also includes the obligation to make DVI Technology components available for board designs which will allow other manufacturers to build the technology at any level. If the developer of the technology tried to keep it all to himself, market growth would be inhibited because no large market will allow itself to get in a position where it depends on only one manufacturer of key hardware.

Software standardization is a different matter, because it runs into the need for data and programs to be exchanged between platforms. This is the field where independent standardizing groups are most important. However, even here, the manufacturer cannot simply hand over the task to others; for example, there is no independent body who will take over the task of standardizing the programming interface for DVI Technology at this time. Intel has to spearhead that themselves.

Another reason why a technology developer has to do much of the work himself is the time scale of standardizing. Standards bodies are for the most part voluntary groups — most of the people have other jobs which take first priority on their time. A standards project does not get the around-the-clock attention that a technology development project sometimes requires to stay on schedule. However, a technology developer cannot afford to make his investment in a technology and then wait for standards before bringing the technology to market. He must go ahead and exploit any lead over potential competitors in order to best recoup his investment. Thus it is common for a technology to be well established in the market before standardizing gets very far. A technology manufacturer in this situation has the obligation to do the best he can to protect any investment his customers have made in the technology while at the same time having to bend somewhat in the standards arena to reach the necessary consensus there. It's not an easy task.

DVI Technology is a set of hardware elements, a C language software interface, and some disc file formats which allow powerful digital audio/video capabilities to be added to a number of different personal computer platforms. The best feature of DVI Technology is that algorithms are programmable in microcode, enabling us to write new routines for new algorithms and thus track whatever happens in the standardizing world. Intel's marketing philosophy for DVI Technology is that the technology should be applied widely — in as many computing environments as possible. The philosophy further envisions design of dedicated DVI player hardware, particularly for the consumer market. However, the proliferation of the technology to different platforms must be done in a way that will provide the maximum amount of commonality for software developers so that applications developed for one platform can be reasonably ported to the full range of platforms. That means that there must be standardization of software interfaces and of video and audio data files. This will ensure a degree of uniformity about the audio and video portions of an application, but of course it cannot deal with already existing differences in operating system interfaces between platforms.

At the same time, standardization should be done in such a way that it places the fewest boxes around the performance of the system — as much as possible, the door should be left open for improvement of algorithms or other software innovation and for future enhancements of the hardware. Obviously, that is a difficult task for any standard-making body, but it is an objective worth striving for. At this writing, the standards task is in an early stage, and the exact outcome cannot be predicted. In this book, we will present the current state of the DVI software developed for the PS/2 and PC AT platforms, which, although it is a mature design, must be viewed as a starting point for standards purposes. However, it is reasonable to expect that most of the existing concepts will survive into the final standards. As already stated, the DVI runtime software provides a C language application programmer's interface (API) for application development. That interface is one of the principal elements of a DVI standard — porting DVI Technology to a different hardware platform will have to maintain the same C language

API. This will allow the DVI parts of an application to be readily ported between platforms.

With respect to standardization of CD-ROM usage for DVI Technology, a standard format for DVI audio/video files must be used on all platforms and on all different storage media. This is the AVSS format, which will be covered in Chapter 12. Because AVSS files use compressed data, they move into a broader standardizing arena. Audio and video compression are used in video teleconferencing, advanced TV broadcasting, and other telecommunications systems. There is a large effort underway to develop international standards for compression algorithms — we will cover that in Chapter 12 after we have discussed compression technology in detail.

Program code may also be included on CD-ROM discs for DVI, and a standard disc will have to include separate program code for each platform that it is expected to run on. All the files on DVI CD-ROMs are standard DOS binary or text files accessed under the ISO 9660 file standard.

DVI hardware will be described in more detail in the next chapter, and DVI software will be described in Chapter 9.

8

DVI Hardware

In this chapter we will be looking in detail at the DVI Technology hardware which you got a glimpse of in the previous chapter. For those readers who have a special interest in hardware, we will depart a little from our stated charter by covering more about the hardware than just what is needed to understand the rest of the book. Specifically, we will discuss the embodiment of hardware used in the ActionMedia™ 750 boards. The heart of the DVI Technology hardware is of course the i750™ chip set — those chips will be used universally in all versions of DVI hardware built by DVI hardware licensees. Therefore, let's look more closely at the i750 chips, beginning with their history.

Part of the original DVI concept was that the video processing should be entirely software controlled, and therefore a custom video processing engine and display device (the chip set — originally called the VDP for Video Display Processor) would be needed to carry out the software-based video features. It was believed that this could be done with VLSI technology, and as that technology developed, the VDP would be low in cost and could even become part of a low-priced consumer product. It was further understood that there would be several steps of design to lead to the final low-cost configuration, and therefore the first design was done using methodology which contributed to achieving a design accurately and fast, rather than producing the ultimate lowest cost chips. Later design iterations would bring the production cost down.

Upon developing a preliminary architecture for the VDP, several other points became clear. The VDP was going to contain so much circuitry that the traditional step of building a breadboard with discrete circuits before attempting full design of a VLSI chip would actually take more time and would cost more than the complete design of a chip from scratch. This led to the search for a design technique for the actual chip which would be certain enough that the breadboarding step could be skipped. The technology of *silicon compilation* (SC) was seen as the answer. True, SC made a chip considerably larger (and therefore more costly to fabricate) than traditional handcrafted design of a custom chip, but it offered much lower design cost, faster design, and a good bag of design verification tools. With considerable controversy, the decision was made to proceed on an SC design of the VDP without any breadboard preceding it. In hindsight, it was the right choice.

As the chip architecture developed, it became clear that the chip would have too many I/O pins for a practical single chip package. (Pin count on VLSI is a significant element of cost.) The VDP would have to be split into two or more chips — thus becoming a *chip set*. The division between pixel processing functions and displaying functions turned out to be a way to split the architecture in two and not to have to add too many pins for the interfaces. Even so, it would take two chips with more than 100 pins each, and with the SC design both of them would be close to the limits of current fabrication capability (in 1986). However, it was decided to go ahead and complete the design, knowing that the rapid progress of VLSI technology in cost and capability would enable the final goals to be met in time. That has proved to be a good choice, too.

Without continuing to relate the design saga (which itself could deserve a whole book), it is sufficient to say that the design of the first VDP was begun and successfully completed within 1986, and the first silicon came available and worked at the end of that year. In parallel with that work, boards and software were designed so that the chips could be tested in a system based on a PC AT. That was another good choice, and the first system actually worked and played motion video from a CD-ROM! That original VDP chip set was custom re-designed by Intel and is now the i750 product.

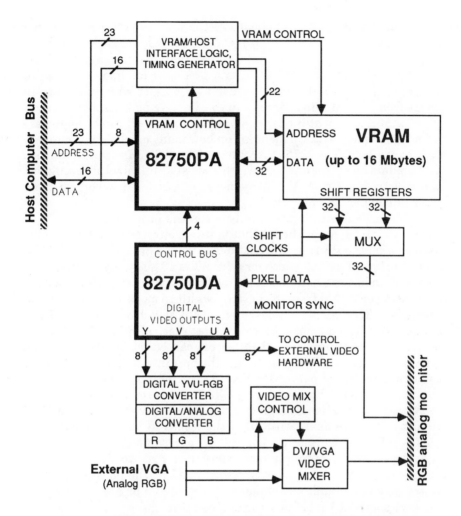

Figure 8.1   Block diagram of the video portion of the ActionMedia Delivery Board showing interfacing of the i750 chips

Interfacing the i750 Chip Set

The internal structure of the i750 chips at this time is only available under non-disclosure from Intel and will not be discussed here. However, we will talk about the characteristics and performance of the chip set as seen from outside and about how the chips are interfaced to each other and to the host computer system. Figure 8.1 is a block diagram of how the i750 chip set interfaces to VRAM, color monitor, and CPU. As already explained, the 82750 PA is a processor for images stored in VRAM, and the 82750 DA is the display processor which converts an image from VRAM into a signal suitable for driving a display monitor.

VRAM is shown at the top right in Figure 8.1 — it can be anywhere from 1 to 16 megabytes, depending on how many of what type memory devices are used. The 82750 PA connects to the VRAM parallel port with a 32-bit data bus. This data bus width allows one full pixel at the highest bpp to be accessed in one VRAM memory cycle, and with lower bpp, more than 1 pixel per VRAM cycle.

There is a separate interface between the host computer bus and VRAM which bypasses the PA (but is still arbitrated by the PA). This path around the PA from the host bus is for the host to access VRAM. Since the host data bus width for a PC AT is 16 bits, only 16 bits of the VRAM data bus are handled at a time for host accesses. However, the addressing of VRAM from the host allows all 16 megabytes of VRAM to be reached. This interface is designed to bank switch 64 kbytes of VRAM at a time into host memory space in order to reach all of VRAM. The host memory segment used for VRAM access can be set by hardware and software switches. Since the PA and the host interface are sharing the same random-access port on the VRAM chips, only one of them at a time can actually be reading or writing VRAM. This requires some care in software design to handle data transfers efficiently. However, the DVI runtime software deals with all of the issues in this paragraph and makes them invisible to an application programmer — through the runtime software, a programmer can read or write to VRAM with 16-megabyte linear VRAM address values, and he/she does not need to worry about the bank switching that is going on or about what host memory segment VRAM is being mapped through.

The control interface to the 82750 PA is directly from the host bus through 256 16-bit registers on the PA. The video board also has an interrupt interface to the system bus. This is used so that operations on either the PA or the DA can interrupt the host CPU. Interrupts generated by the DA are transmitted to the host via the PA, so there is need for only one interrupt interface servicing both chips.

The 82750 PA is a microprocessor with its own instruction set and needs a program for running those instructions. As we have already explained, programs for the PA are in microcode — executed from an on-chip RAM. Microcode gets to the PA by first being loaded by the host system into VRAM, and then during execution the PA moves it into its own RAM. The DVI runtime software normally manages the transfer of microcode, either from host to VRAM or from VRAM to the PA. An application programmer never needs to worry about microcode.

The 82750 DA does not directly access the host system bus, rather it connects to the PA and the VRAM serial port. In order to handle the extreme data rates needed for the color monitor refresh, the serial outputs of the VRAM chips are arranged in two 32-bit buses, connected by a multiplexer (MUX). The output of the multiplexer is a single 32-bit data bus which goes into the DA where the pixel format of memory is decoded into a standard 32-bit output format.

The DVI delivery board is designed for the DA to operate with VRAM pixel formats based on luminance/chrominance components(Y-C) rather than RGB. This allows the same advantages that YIQ brings to NTSC television to be exploited in a computer video system. (In general, these advantages can be obtained by using any format which separates the colors into luminance and chrominance components. The DVI hardware could be operated in any Y-C format, such as YUV; Y,R-Y,B-Y; etc.) the 82750 DA also contains special hardware which takes care of some of the processing necessary when displaying images that are stored in luminance-chrominance format. We will discuss the Y-C format in more detail in Chapter 11. The ActionMedia boards are designed to operate with YVU — this is the DVI standard format.

The 82750 DA delivers a 32-bit parallel output which contains 8 bits each of luminance and two color difference components plus an 8-bit alpha-channel value which can be used off-chip for various purposes — for example, controlling an external video mixer. (The current delivery board does not include such a video mixer.) The Y-C digital outputs are processed from YVU to RGB and then sent to a triple D/A converter chip on the delivery board for conversion to analog. Then they pass through an analog video matrix to convert to RGB. The use of Y-C as the pixel format both in VRAM and at the output of the DA is a software choice, but of course the hardware must contain the correct matrix circuit to convert from the format used at the output of the DA to the RGB signals needed for a standard monitor. (The DA contains some hardware features that are specific to the use of luminance-chrominance formats; but if the DA output format was programmed to be RGB, those special features of the DA could not be used.)

The delivery board allows an external VGA to be combined with the

DVI RGB video by means of an analog video mixer contained on the board. The mixer may be controlled from either the VGA or DVI outputs to select either source for display, or to switch between them on an individual pixel basis.

The 82750 DA is responsible for the scanning sync frequencies and the sync pulse formats of the output. Control registers on the DA are loaded with the values for any desired sync configuration. This is normally handled by the runtime software.

The DA talks to the PA through the 4-bit VBUS, which is a unidirectional bus used only for control purposes. In order to set up the many control registers on the DA, the information is first loaded into VRAM by the host CPU. Then the host tells the PA to load the DA. The reason the PA gets the command is that only the PA has access to the addressing of VRAM. In order to load its registers, the DA asks the PA to initiate a register load. The PA immediately responds by commanding VRAM to load the register table into the VRAM's shift registers. Since the DA does have access to the VRAM shift register outputs, subsequent clocking of the shift registers will then transfer the register values into the DA. Again this is a sequence that the DVI runtime software handles for the programmer automatically as it is required.

82750 DA Functionality

Because it totally controls the output display, the 82750 DA must be set up for the exact display characteristics required by the application. With the DA's flexibility, this setup is quite complex; but the DVI runtime software has functions which will simplify that for the programmer in most cases. To give a better understanding here of what can actually be done with the DA, the following discussion looks at what could be controllable if you access the DA chip directly. In that respect, it may be too detailed for some readers — you may skip over this section if you wish, because the simpler methods of programming the DA will be covered later in Chapter 9. However, those readers wishing to gain a fuller understanding of the flexibility of the 82750 DA will find the following to be useful.

All the parameters of the DA are determined by control registers on-chip. Loading of the control registers is done by the DA itself, which moves values from a block of VRAM which is allocated by the runtime software for this purpose. The actual move from VRAM occurs during

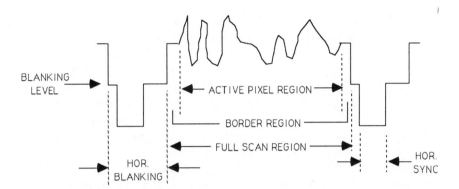

Figure 8.2   **Video line waveform showing programming points for the 82750 Display Processor**

the nonactive video time of the display signals so that the video output is not interrupted. Thus the chip set's registers can be changed while an image is being displayed. Loading of the 82750 DA registers from VRAM is managed by the DVI runtime software. Characteristics programmed via the 82750 DA's registers include:

- Resolution
- Active pixel area
- Video and sync formats
- Screen border
- Pixel format
- Interrupt generation
- Color map entries

Programming horizontal frequency and resolution
The 82750 DA has its own clock frequency determined by a crystal located on board. The crystal frequency must be chosen based on the resolutions and screen formats desired and on any need for compatibility with another standard, such as NTSC or PAL. The highest clock frequency that can be used is 22.5 MHz. Much of the programming of the DA is done in terms of the period of the DA's clock, which is called a *T-cycle*. In the example which follows we will use a 20.00 MHz clock, so one T-cycle is exactly 50 nanoseconds.

The pixel rates which the DA can generate are divided down from the DA's clock frequency by a factor chosen from this list:

$$1.0, 1.5, 2.0, 2.5, 3.0, 3.5, 4.0, 4.5, 5.0, 5.5, 6.0, 6.5, 7.0$$

providing 13 different values for pixel rate. In our example, we could thus choose pixel rates from 20 million pixels per second down to 20/7 million pixels per second. The lower pixel rates of course would be used for lower resolution display formats.

In order to determine horizontal resolution, three other parameters are involved — the line scanning frequency, the horizontal blanking interval, and the horizontal active pixel area. This is shown graphically in Figure 8.2. Line frequency is determined by specifying how many T-cycles are in a half-line period. For example, to get as close as possible to the NTSC horizontal frequency (15,734 Hz) which has a half-line period of 31.78 microseconds, we would use 635 T-cycles for a half-line. This example is the first line in Table 8.1. With the 20 MHz 82750 DA clock, this would give a horizontal frequency of 15,748 Hz. (For exact matching of NTSC sync, one would choose a clock frequency slightly different from 20.00 MHz so as to make line frequency come out precisely at 15734 Hz.)

With the half-line divisor of 635, a line can contain a maximum of 635x2=1270 pixels, at the lowest available pixel rate divisor of 1.0. However, this number also includes pixel cycles during the horizontal blanking time. If we wanted the full screen width to be represented by 1024 pixels (the largest number the chip set will support), then we would set up the DA to devote 246 T-cycles (1270–1024) to the horizontal blanking interval.

The *active pixel region*, which is the part of the full screen width that is actually taken from the VRAM bitmap being displayed, is programmed separately. For computer display monitors which are normally *underscanned* (with the edges of the scan visible), the active pixel region would normally be equal to the full-screen pixel width. (That is what we did in the 1024-pixel example above.) However, television display monitors are usually *overscanned* so that the horizontal and vertical scanning of the CRT goes beyond the edges of the display. This is done so that the edges of the raster are not visible — for television viewing the picture edges are considered to be distracting to the program content. Of course, for a computer display being shown on a TV, particularly a computer screen that includes text, we would not want the

Table 8.1 — Programming of the 82750 DA for Different Formats

Resolution	Inter-lace	pixels /line	Hor blank	Pixel divisor	Line divisor	Vert divisor	Vert blank
1024x480	yes	1270	246	1.0	635	525	45
512x480	no	634	122	1.0	317	1050	90
640x350	no	832	192	1.0	416	800	100
320x200	no	404	84	3.5	707	470	70
256x240	yes	318	256	4.0	636	525	45

(DA clock=20.00 MHz, active pixels=full scan, vert=60 Hz)

text to go off the edges of the display. Therefore, for this type of use, the active region is specified smaller (usually by about 10%) than the full scanning width such that all of it will be visible on an overscanned TV. When the active pixel region is specified this way, the pixels outside of the active pixel region are called the *border*. The 82750 DA can separately program the border pixels to have any color specified by a 24-bit format.

Programming vertical frequency and resolution
The vertical scanning is specified to the DA in terms of the number of half-lines per scanning field. If an odd number of half-lines is given, we will get an interlaced scan. For example, for NTSC vertical scanning (first line of Table 8.1), there would be 525 half-lines per field, and it would take two fields to display a full frame. If the number of half-lines per field is even, then we get noninterlaced scanning. To provide a 525-line noninterlaced scan, we would specify 1050 half-lines per field as shown in the second line of Table 8.1. Note, however, that if we still used the horizontal line frequency we set up earlier, we will get 525-line non-interlaced scanning at a vertical scanning rate of 30 Hz, which will flicker terribly. To get a noninterlaced scan of 525 lines at 60 Hz vertical rate, we have to go back and change the line frequency divisor to be half the value used earlier. But this will drop the maximum number of pixels per line in half, so the highest number of horizontal pixels per full-scan in a non-interlaced 525-line scan is 512 pixels. (Higher horizontal resolution is possible in a noninterlaced system if we use fewer than 525 lines per frame — see line 3 of Table 8.1.)

The vertical blanking period and the vertical active region also must be specified in half-line steps. These parameters work just like their horizontal equivalents. If we specified vertical blanking to be 45 half-lines for the interlaced-scan example and the vertical active region to be full-screen, then we would have set up a display of 1024x480 interlaced display. Similarly, for the noninterlaced example, we could set vertical blanking to 90 half-lines and the vertical active region to full-screen to get a 512x480 noninterlaced display. You can see that the DA provides a lot of flexibility for setting up different scanning numbers, however, the maximum resolution numbers are limited by the highest pixel rate that can be supported, which is equal to the DA clock frequency.

The 82750 DA has additional registers which allow separately setting up the start and stop timings for horizontal and vertical synchronizing pulses. These also are programmed in steps of T-cycles and half-lines. In the case of vertical sync waveforms, the format of NTSC vertical sync is fully supported. (See Appendix A for NTSC sync waveform.)

Having such tremendous flexibility in the hardware has advantages and disadvantages. The advantage is that you can match almost any set of numbers within the limit of pixel rate. A disadvantage is that if the hardware is not properly programmed, it can put out very crazy scanning frequencies and sync waveforms. In some cases, this might actually damage a display monitor, particularly one of the multiple-sync variety. It is desirable that the supporting hardware for the DA and the system software be designed to include some kind of protection against damage by wrong sync—this is particularly important for hardware that is used during development of the DA software. Similarly, software design for the DA should try to minimize the possibility of incorrect setup of the DA.

A second disadvantage of the tremendous flexibility of the 82750 DA is that the programming of it is complex and tricky. Not everyone would really like to be worrying about all that. This is where the DVI runtime software comes in. The DVI runtime software has simple subroutines which will set up the DA for you, and you only need to worry about the numbers you want, and not about how to achieve them. A reasonable subset of the DA's capability is available this way, which should satisfy most applications.

But we have only talked about part of the programming of the 82750 DA. We also have to tell the DA about the bitmap we want to display. The DA can display a bitmap located to start on any 2 kbyte boundary in VRAM. There is also a restriction that the pitch of a display bitmap must be an exact power of 2 (128, 256, 512, etc.). However, any width of bitmap can be displayed simply by setting the pitch to the next highest power of 2. (For example, to display a bitmap with a width of 320 pixels, the bitmap pitch would have to be 512 pixels.) The DA already knows about the resolution values (active pixels and lines) set up in the paragraphs above. In addition to the address, the DA must be told about the bpp format of the bitmap and the type of processing to use in displaying the chrominance part of the bitmap. Typical bits-per-pixel formats are as follows:

- 8 bpp, with 24-bit CLUT
- Special video mode — 9 bpp
- Mixed video/CLUT mode — 9 bpp
- 16 bpp — YVU
- 32 bpp — YVU plus Alpha

A further area of the 82750 DA comes into play when color-map modes are used. These modes are selected by the DA registers, and two methods of mixing color-map and real-video modes in the same bitmap are selectable by other DA registers. The color map values themselves are also held in DA registers. The pixel formats will be described in more detail as part of the software examples in following chapters.

The 82750 DA has control registers for setting up the digitizing mode for input of a digitized image into VRAM (explained below), and also registers to control interrupt of the host CPU at any desired point in the scanning pattern. This latter feature allows more than one DA mode to be used in the same image, as well as any other activity which would need to be synchronized with the image scanning.

The 82750 DA also has provision for use of an alpha channel in all bpp modes. The alpha channel is a separate 8-bit parallel output from the DA that runs at the pixel rate. One byte is output for every pixel. There are several registers that control the operation of the alpha channel, including a scheme where the alpha channel output is the result of comparing an 8-bit value in an Alpha Trap register on the DA with the Y value of each pixel. The alpha channel then outputs one of two 8-bit values, depending on whether the compare is true or false. This is a way to create in real time a separate keying or control signal that depends on the Y-value of the pixels in an image.

In the 32-bpp mode, the alpha channel outputs the 8 highest bits of each pixel. In this case, the alpha byte can represent a separate image for any purpose desired. (The alpha channel could also be used as a separate 8-bit image plane, with appropriate combining hardware added off-chip.) Note that the current Delivery Board does not have any hardware for the alpha channel. And, of course, the functionality just described exists in all 82750 DAs.

82750 PA Functionality

The pixel-processing functionality of the 82750 PA is controlled by its software. Many parts of the PA resemble a traditional microprocessor, in fact it could be used as the main CPU in a personal computer. However, it is programmed in microcode, and it would prove awkward to make an entire PC programmed in microcode. But microcode is an ideal approach for programming the functions we need for pixel processing, and that's why it is used in the 82750 PA. In this book we will not be describing the details of PA microcode; however, there are some general things that can be said.

The 82750 PA has some of the generic elements of pixel processing built into its hardware. These are accessible in the form of pixel-processing instructions in the PA's instruction set; such instructions facilitate doing complex pixel manipulations efficiently, with fewer instructions per pixel than a typical general-purpose processor would require. The PA also has a lot of parallelism — for example, there are loop counters built in to do the housekeeping for the two nested loops that are used often in two-dimensional bitmap operations. The loop counters operate in parallel with the rest of the chip, so no instruction cycles are used just for counting loops. It has similar parallel features for accessing sequential locations in VRAM — another thing used often with images. Every instruction in the PA runs in one PA clock cycle — 80 nanoseconds for a 12.5 MHz clock (which is the normal PA clock frequency). Since VRAM is much slower than 80 nanoseconds, the PA's data bus to VRAM is 32 bits wide to speed up data moves to and from VRAM.

To demonstrate the 82750 PA's speed in performing complex pixel operations, the *warp* algorithm is a good example. Warp performs the task of mapping one polygonal region into another polygonal region having the same number of vertices. Warp is used, for example, to fill all the polygons in a perspective view of a computer model using video textured surfaces to create a realistic rendering of a 3-D object or scene. Figure 8.3 shows an example of the use of warp to fill a diamond-shaped

Figure 8.3   An image of some tulips is transformed into four diamond-shaped figures by use of the warp algorithm

object with a fabric texture which is stored as a rectangle. On the PA the warp algorithm runs at 500,000 pixels per second. Simpler graphics operations run faster, for example, filling rectangles or polygons with solid colors runs at 4–8 million pixels per second, depending on bpp. The speed potential of the 82750 PA is really put to the test in performing complex image compression or decompression algorithms — this will be discussed in Chapters 11 and 12. Intel is designing faster versions of the PA for future introduction. This will speed up operation by a factor of 2 or more while maintaining software compatibility with the present chips.

Microcode is handled by the DVI runtime software in a manner similar to the 82750 DA register contents. It is first moved into VRAM by the runtime software, and then the 82750 PA reads it from there into its internal RAM for execution. A section of VRAM is allocated by the application program to hold all the microcode routines for an application; then the routines are loaded from mass storage into VRAM the first time they are called. A brief delay caused by disk access occurs the first time each different microcode routine is called in an application. Subsequent calls will simply reuse the microcode already in VRAM, and there will be no delay.

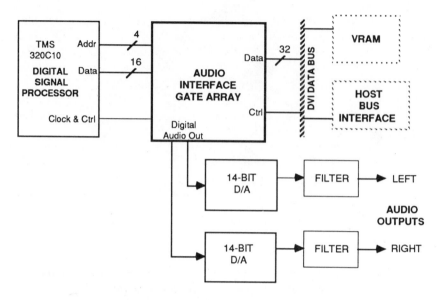

Figure 8.4   Diagram of the ActionMedia Delivery Board audio circuits

DVI audio processor hardware

The DVI standards approach specifies the audio system at a software level. There will not be a standard audio hardware for DVI Technology. In principle, different audio hardware can be used, with the appropriate software drivers to bring that hardware capability out to the standard software interface. Here we will describe the hardware that is included on the ActionMedia 750 Delivery Board. Figure 8.4 is a block diagram of the audio circuits on that board.

The TMS320C10, a CMOS digital signal processing (DSP) chip, is the audio coprocessor. This chip, made by Texas Instruments, uses on-chip ROM to hold its programs. The DSP programs are written in the chip's own assembly language, which uses a 16-bit instruction format. All except a few DSP instructions are one word long, and they execute in one DSP CPU cycle, which is 1/4 of the DSP clock frequency. A 32 MHz clock frequency is used, which gives an instruction rate of 8 MIPS, or one instruction executed every 125 nanoseconds. The ActionMedia audio processor software include several algorithms which can be selected by host software.

All interfaces between the DSP and the rest of the system are handled by a custom gate array on the ActionMedia board. Audio data for the audio processor comes from the host system via VRAM. A buffer is allocated in VRAM to hold up to 4 seconds of compressed audio data. This allows the audio data accesses from hard disk or CD-ROM to be scheduled independent of the constant need to pass audio data to the coprocessor. That also contributes to more efficient utilization of the host computer's bus bandwidth.

Audio output is delivered via the gate array to a pair of 14-bit D/A converters for conversion to analog. The converters are followed by programmable-bandwidth analog filters, which remove any D/A artifacts from the output. The two channels may be used as a stereo audio channel, or they can be used separately for two different functions — the channel usage is controlled by software, of course.

ActionMedia 750 capture board

The Capture Board, shown in Figure 8.5, allows digitizing of video still images or motion video, and capture of digital audio, either with video or separately. Two-channel or stereo audio is first processed by programmable analog chips to set levels and filtering and is then digitized at 14-bits by two A/D converters. The output from the A/Ds is serial, and a gate array chip converts them to a parallel format suitable for passing to the DSP on the Delivery Board for compression. The resulting compressed audio is stored on hard disk.

The video digitizer has an RGB video input. (To digitize an NTSC composite video input, an external NTSC-RGB decoder must be used.) The input is first converted to YUV format with an analog matrix. Then analog filtering is applied to remove any frequencies above half the sampling rate, and the components are digitized by three A/D converters. The digitizing resolution is 768 x 480, and the digital image is placed in VRAM on the Delivery Board. Horizontal and vertical synchronizing signals from the input are extracted and passed to the Delivery Board for genlocking purposes.

For still images, the digitized output from VRAM can be stored on hard disk in a standard DVI image file. There also is a *scaling* algorithm that runs on the 82750 PA for changing the size of an image, so the digitized image can be reduced or enlarged by any amount, cropped, and saved to a standard DVI image file at any pixel resolution desired. Note that DVI still images do not have to be digitized with the digitizer described here. They can be captured on other equipment, and their file

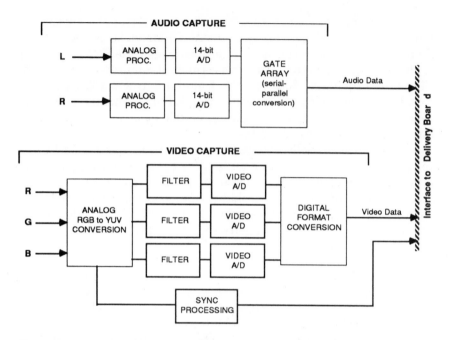

Figure 8.5   Block diagram of the ActionMedia 750 Capture Board

format converted to a DVI image file format. This is a way of using already-existing digitized images.

During digitizing, the 82750 PA is idle, and therefore it can be programmed to do compression of motion images as they are captured. One example of this is the DVI *Real-Time Video* (RTV) software, which sets up the 82750 PA to perform a compression algorithm on motion images as they are digitized by the video digitizer. RTV allows capture of motion video at real time on a DVI system and stores the result on hard disk. Because a simplified compression algorithm has to be used, the pictures reproduced by RTV have a frame rate reduced to about 10 frames per second when compressing to CD-ROM data rates , and they have more artifacts than the full-quality DVI compressed video, which requires the use of a large computing facility. However, for slightly higher data rates such as can be provided by hard disks, there are other options in RTV that operate at 15 frames per second or even 30 frames per second. This is a good example of DVI Technology's ability to change algorithms through software programming.

Summary

The ActionMedia 750 boards just described were designed to support DVI Technology application development and delivery on familiar platforms, the IBM PS/2 and PC AT. These boards are cost-competitive with other high-resolution video/graphics boards even though they also include DVI Technology's unique motion video and audio capabilities. Intel plans to cost-reduce the ActionMedia boards as well as to license hardware companies to build DVI capability into their platforms or to create new platforms with DVI Technology built in. The forecast is that a dedicated DVI system will become available in the mid 1990s in the $1000 price range, which is necessary for broad use in homes and schools.

9

DVI Software

So far, we have deferred much of the discussion about the details of DVI Technology's capabilities, hiding behind the statement, "It's all in the software." Well, in this chapter, we won't hide any more, because here we really will talk about the DVI software. In particular, this chapter will give an overview of the DVI software for the ActionMedia™ 750 family of boards, presenting the architecture and discussing some of the reasons for the particular choices which have been made. The chapter ends with an example of a DVI graphics program.

The DVI software discussed here is the current release at the time of writing (June 1990), and it is the second-generation design. The architecture is well thought out and provides a lot of functionality, yet it is easy to use and has capability for enhancement and expansions.

The software ties the DVI hardware together and makes it into a true multimedia system having all types of visual and aural presentation simultaneously available. As we have said before, *too much* flexibility can become a burden in that the system may get complex and difficult to program. This issue has been carefully considered in the design of the DVI software, and what has resulted is a package of software which provides a C-language programmer access to DVI Technology's capabilities at two different levels:

- A high-level interface where the programmer does not have to worry about details beyond those needed for adequate flexibility of function, and
- A lower-level interface where the programmer can get closer to the hardware to do special things, but he/she has to be concerned with details.

Both of these interfaces are part of the DVI Video Application Programming Interface (VAPI), and DVI software designed for different platforms will try to maintain these same interfaces. This will aid portability of the DVI part of application code between different DVI platforms.

There is also the possibility of having still higher level interfaces to DVI Technology which would not require any C programming, and might even be a real-time interpretive interface, which would execute commands immediately without the need to compile and link. Several programs of this type were started during the development of DVI Technology, and one of those is included with the DVI developer's software package. That is the **vgrcmd** interpreter, which allows working with all of the DVI graphics commands from the keyboard in real time. It is not higher level in the sense that the command structure is simpler to use in performing complex tasks, but it is interpreted, and it is useful for trying out graphics sequences before writing them into a C program.

The ultimate high level programming tool is what is called an *authoring* language or system. Many authoring languages exist for other multimedia systems such as those which combine a laser disc with a computer. With an authoring language, the structure of an application can be constructed in either flow chart or storyboard form, and all of the different screens and sequences can be described in simple language or with a point-and-click graphical interface. Then the authoring software will create the application code and let the developer run the application and interactively experiment with the results to make improvements or modifications. When this process (authoring) is completed, a finished application is produced. No knowledge is necessary of C language or of many of the internal system details. Several products of this type are on the market for DVI Technology — they will be described in Chapter 14.

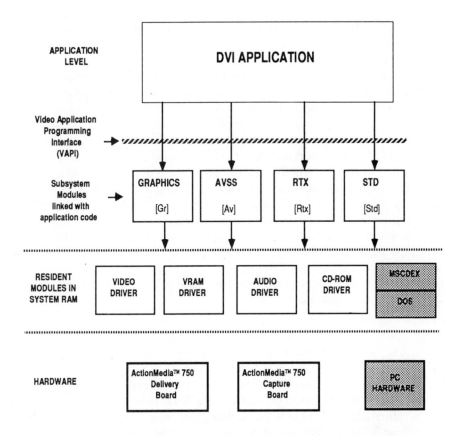

Figure 9.1   DVI runtime software architecture showing standard interfaces. The shaded boxes are standard PC hardware and software; the clear boxes are the parts of the DVI hardware and software.

Figure 9.1 is a block diagram of the DVI runtime software showing the different modules and their relationship to the hardware and to each other. The hardware (at the bottom of the figure) is talked to directly by *drivers* which are installed at boot-up and remain resident thereafter. (A driver is a software module which provides interface to a particular piece of hardware.) There are four drivers, dealing with interface to VRAM, the i750 chips (video), audio, and the CD-ROM interface. These drivers are not accessible to application programs except through the subsystem modules which sit above the drivers.   The rest of the DVI runtime software is contained in the subsystem modules, which become part of an application program through linking.

The driver modules create *virtual devices* for use by the higher-level software.   A virtual device is a software entity which defines the interface characteristics of a physical device, such as a video display. The virtual device description may differ from the actual physical device in that the virtual device software may deal with device operating details such as interrupts or port interfaces, hiding these complexities from the higher-level software which accesses the virtual device interface.

Before we go any further, we should talk about the naming conventions that are used in the DVI runtime software. All interfaces by application programs are through C-function calls. These functions are named with a prefix which indicates which subsystem module they are in. These naming prefixes are given in Figure 9.1 in brackets, as they are below in the text.

There are four subsystem modules, as follows (the function name prefix is given in brackets):

- Graphics [Gr] — The package which provides image handling, graphics drawing primitives, and video manipulation functions. Detailed description of the graphics calls will begin below.
- AVSS [Av] — Audio/Video Support System — This package is used for playback of audio/video files that are in the AVSS format. It can play audio and/or video from CD-ROM, hard disk, or RAM. AVSS will be further described in Chapter 12.
- RTX [Rtx] — Real-Time eXecutive— Provides multitasking inside of a DVI application. All DVI applications use RTX which will be described in Chapters 12 and 14.
- STD [Std] — A package of functions for memory management, I/O, and error management.

Another important module is a DOS extension provided by Microsoft:

- MSCDEX — Microsoft CD-ROM Extension — this extends DOS so that files on ISO 9660 format CD-ROMs may be accessed by DOS in exactly the same way files are accessed on floppy or hard disks.

Figure 9.2   Examples of the three graphics contexts — images, structured graphics, and text

The DVI Graphics Model

The DVI Graphics package contains a group of routines which manipulate information in bitmaps. A bitmap anywhere in VRAM may be operated on by the graphics routines, even the bitmap which is currently being displayed by the 82750 DA (the *screen* bitmap). The graphics routines can operate in three different contexts:

- structured graphics — the destination bitmap is written into by using *primitive* functions (lines, rectangles, polygons, etc.); only the pixels involved in the particular object are affected by the primitive function,
- images — source and destination rectangles are defined; during an image operation all pixels are processed the same way,
- text— the destination bitmap is operated on in a rectangular format which represents one text character at a time, using character description data taken from a separate font storage area.

Figure 9.2 is a DVI screen photograph which shows examples of each of these types of graphics operation.

Structured graphics

Structured graphics is widely used on personal computers, and there are a number of industry standard systems in use, such as GDI in Microsoft Windows, Digital Research's GEM, Apple Macintosh's Quickdraw, and others. These standards are intended to allow interchange of structured graphics command data between different applications on different machines, using files of data which the system can read to draw graphics screens of any complexity. However, the different systems are not compatible with each other. The DVI software must solve the same problem — there needs to be a standard format for DVI graphics data so that different DVI systems can share that data.

Some of the industry standards go a step further in that they try to work at a level which will be *device independent*. That means that the graphics data is stored in a general format which is not specific to any of the display device characteristics such as resolution numbers, pixel formats, etc. The display hardware and software drivers (which will know about the exact characteristics of the display) will convert the general data into the commands needed for that display to come as close as possible to what the standard file is specifying. This conversion of course has to be done in real time as graphics are displayed — the result is that such systems can be slow if their CPU or their graphics hardware is not powerful enough.

The DVI structured graphics environment is quite different from a device independent system — instead of hiding the device characteristics behind a standard interface, DVI Technology lets the programmer *specify* the device (through the calls which program the 82750 DA) to have exactly the characteristics which are optimum for the application at hand. Parameters such as resolutions, bits per pixel, and scanning formats are specifiable by the programmer to suit the specific needs of the application. The programmer then works with this optimized device through the DVI graphics function calls. The result is a very flexible and, at the same time, a very fast system.

Images

The DVI Graphics package has a number of routines for operating on rectangular blocks of pixels — these are the image manipulation routines. The key characteristic of an image routine is that all the pixels in a block receive exactly the same treatment, whether it is simply a copy to a new location, or an elaborate image-processing function. This kind of function is not common in personal computers today, and there is much less existing standardization to consider. DVI Technology has defined file formats for images which provide for all of the flexibility of DVITechnology to be utilized and for future expansion to new functionality. We will look at the image functions and the image file format in some detail in Chapter 10.

Text

The text functions in the DVI Graphics package use *bitmap fonts* as the source to place text characters at any location in a bitmap. A bitmap font is simply an image which contains pixel data for all the characters along with a table of coordinates and sizes for the characters. In the present DVI system, the bitmap fonts can be 1 bit per pixel or 2 bits per pixel, although the font file structure will allow future addition of other formats. There are other kinds of fonts which can provide additional features. One type is the *vector* or *stroke* font, where the font file contains only a description of the outline for each character expressed mathematically as a series of lines and/or curves. The font-handling software must convert this description into actual bitmap characters in real time at the time of drawing text. The advantage here is that characters can be drawn in any size or almost any style from a single font. The disadvantage, of course, is that drawing of text takes much more processing power. (The power of the 82750 PA is appropriate for doing vector fonts very rapidly, but so far no one has written microcode for it.) The text routines support management of a number of fonts, and they handle the housekeeping of finding each character in the font data and copying that into the destination bitmap.

Graphics Data Structures

There is a fundamental structure, defined in the DVI graphics package, which are used by all the DVI video software: the *bitmap data structure*. A bitmap data structure tells the runtime software about a bitmap.

A data structure is set up for each bitmap that an application will use. Any number of bitmaps can be specified, even reusing the same locations in VRAM. This software-based layout of VRAM is one of the most valuable flexibility features of DVI Technology because it allows the limited resource of VRAM to be configured optimally for each application, or even dynamically reconfigured during an application. Each bitmap structure contains data fields for bitmap address, size, pitch, bpp, format, and various attributes used by the graphics functions.

Bitmap structures are located in host system RAM, but they are never accessed directly in an application, rather they are reached via special **Get** and **Set** functions. The use of functions to access these structures allows the structures themselves to be extended or modified in future releases of the DVI software, and all the application accesses will still work simply by recompiling with updated function declarations (contained in *header files*, which will be explained shortly) and relinking to updated system libraries. The **Get** and **Set** access approach is used for all DVI data structures. All the graphics functions are passed bitmap *handles* which tell them which bitmap(s) to operate on and how to perform the operation. (A handle is used by the system to find the address of the start of the data structure in RAM.)

Attribute fields in a bitmap structure include:

- DrawColor—tells the primitive functions and the text functions what color to use for drawing an object.
- DrawOutline— tells the Rectangle, Polygon, and Ellipse functions whether to create a filled or outline figure using the DrawColor.
- DrawPlane—tells the Rectangle, Polygon, and Ellipse functions which plane (Y, V, or U) or planes of the bitmap to draw into.
- Rop2 — this field applies to the graphics functions such as Copy and the drawing primitives such as Line, Rectangle, which are capable of doing *raster operations* or *raster-ops*. A raster-op is any logical process which is performed on each pixel by the graphics function. The "2" in the name Rop2 means that the raster-ops are *two-way* — they operate between the original value of each pixel and a second parameter. The type of process is defined by this attribute. Rop2 types are: NONE, OR, AND, EXCLUSIVE-OR, NOT, NEG, INC, DEC, PLUS, MINUS.
- RopType — this field further specifies raster operations for the Copy function which can use several different types of operands. Choices are DST_SRC, which performs the raster-op directly between the destination and source pixel values, or SRC_COLOR, which performs the operation between the source pixel values

and a color specified by the DrawColor field.

- CopyKey—tells the Copy function whether to use transparency or not.

- CopyFilter — tells functions which have a filtering option whether to use it or not.

- CopyKeyColor—(relevant only when CopyKey is TRUE) specifies a color in the source rectangle of a Copy which will appear transparent in the copy (that color will not be copied).

- TextFont — specifies the font which will be used by the Text calls.

- TextGap — specifies the number of pixels between text characters.

- TextBg — specifies whether text is placed with a background color or transparently.

- TextBgColor—when text background is used, this specifies the color.

Additional graphics features in future releases of the DVI software could be supported by expanding the attribute portion of the bitmap data structure to contain additional fields that control the new features. These new fields will be given default values when the bitmap structure is initialized. This will disable the new features so that old software which does not know about the new features will still work. New software would activate the new features by setting the appropriate field values in the bitmap structure.

Display Monitors

The DVI system can support two display monitors at the same time. One is the PC standard output display, which can be monochrome or color, using one of the standard PC display adaptors — MDA, CGA, EGA, or VGA. The other display is an analog RGB monitor which is connected to the ActionMedia 750 delivery board. The first display is used for interface to DOS and is also driven by all the C-language functions such as **printf( )** which affect the standard output device. The second display is the DVI display, controlled by the DVI Technology specific functions. The ActionMedia 750 Delivery Board has the capability to combine both displays on a single VGA monitor, by keying the output from a standard

VGA display adaptor with the DVI output. For situations where a separate command monitor is unnecessary or undesirable, this VGA single-monitor feature is appropriate. Because the VGA and DVI display hardware is separate, with combining occurring at their outputs, you actually gain functionality by using the combined display.

For development of DVI applications, the presence of two monitors is usually an advantage, because the standard display device can be used to show error or status messages created by the application under development without messing up what is on the DVI display. On the other hand, in an application situation, it may sometimes be advantageous to make the standard display device part of the user interface so that video and audio from the DVI display need not be interrupted to present the information needed to operate the application. For the examples that follow, we will assume that there are two monitors present, and we will make use of both of them. (Using the VGA keying feature, both outputs could be combined on the same physical monitor.)

Graphics Routines

At this point we will discuss in detail how the DVI graphics routines are used to perform common tasks. That means we will begin talking in the C language. Anyone who is not familiar with C at least to the level of reading and understanding (you don't necessarily have to be able to program with it) should stop here and review Appendix B before continuing. If you are not into software at all, you should skim across the detail software discussion and then look carefully at the section "Points made by the software example." These sections follow most software examples, and they will recap what was demonstrated by the software example.

Note that the software examples in this book are not a complete exposition of all the DVI software features. Only a subset of functions will be presented and explained. The purpose of these examples is to show you how some common video functions are done with the DVI runtime software and to convince you that it is not difficult.

Intel periodically makes update releases on the DVI developer's software package. The examples here use the current release at the time of writing. Although it is unlikely that any significant changes would be made in the interfaces used here, the final authority on the DVI software you use must be the documentation you receive with that software.

The DVI Development Environment

Programming for DVI Technology is done on an IBM PS/2 or PC AT or compatible computer which contains the DVI ActionMedia 750 Boards and the Developer's Software. This contains:

- DVI Runtime Software libraries
- DVI Runtime Software Include files
- DVI Font files
- DVI Microcode files
- DVI Driver programs
- DVI Authoring Tools

C compiling for DVI Technology is done with the Microsoft C package (the version required is specified in the documentation for the DVI software you are using — here we are using Version 5.1), including the compiler, linker, and (optionally) the Make utility. This development software is normally set up in several directory structures according to the installation notes for both Microsoft C and the DVI Developer's Software. It is not necessary for us to go into that in detail here.

For our example of graphics programming, we will use a simple program **demo.c**, which makes a menu using the graphics and text routines. Several different graphics demonstrations may be selected from this menu. **Demo.c** produces a complete program, which can be run on a DVI development system. (All the functions of this demo program are actually in the same source file **demo.c**, so the header **#include**s and the global variable declarations are not repeated in the listings of each function.) This program also sets up an environment to which other example functions will be added in later chapters simply by adding them to the menu.

The DVI C functions and their usage will be explained as we need them in the program examples. We will try to minimize abstract tutorial discussions in favor of presenting concepts as they are required to carry out the objectives of the example. However, at least one general point will make things easier to follow if we present it at the outset — that is the system of nomenclature for the DVI functions, constants, and variables. Every function or data item name begins with the abbreviation of the software subsystem to which it belongs: graphics names begin with **Gr**, AVSS names begin with **Av**, Rtx names begin with **Rtx**, and standard subsystem names begin with **Std**. The naming continues with a second abbreviation for a sub-subsystem if there is one, and then an action name. So, for example, the graphics function which displays a bitmap is called: **GrBmDisplay( )**, where the abbreviation **Bm** indicates that it is a bitmap-management function, and of course **Display** is obvious. The same naming strategy is also used for variables and constants, except that constants are in all-caps, which means that the upper/lower case way of separating the parts of the name does not work. The underscore_ character is used instead, so a graphics constant which identifies the Y-plane of a bitmap would be GR_PLANE_Y.

This example program will be easier to follow if we cover the overall concepts of the program before we get into the actual source code. (As any programmer knows, this kind of overview is always a good thing to do in planning a program; and it becomes even more important as the complexity of the application increases.) Here are the key features of the program:

- Screen format — The application will use a 512x480 screen at 16 bpp for display. A bitmap for this at address 0x10000 in VRAM will be set up and will always be the display bitmap. (The reason for starting this bitmap at 0x10000 instead of 0x0 is that the first page of VRAM must be reserved for the DVI system to use for audio buffers.) This bitmap is going to take 480 kbytes of VRAM, which goes up to VRAM address 0x87FFF. However, some functions will need additional bitmaps to do off-screen work — and one working bitmap will be defined initially for that purpose. In addition, some functions will need other working bitmaps — they will be set up locally within the function and released after the function is complete. All the working bitmaps will use VRAM addresses above 0x88000.

- Layout of the screen — The menu will be displayed in a 512x36 bar at the top of the screen; it will always be showing. The screen area below the menu bar (512x444) is the display area for the function selected. Functions will be restricted so they will not overwrite the menu.

- Operation of the program — Menu items are selected by the keys on the keyboard. This approach is used because of its simplicity; in Chapter 14 we will modify this same menu for selection using a mouse. The selected function will write to the display area

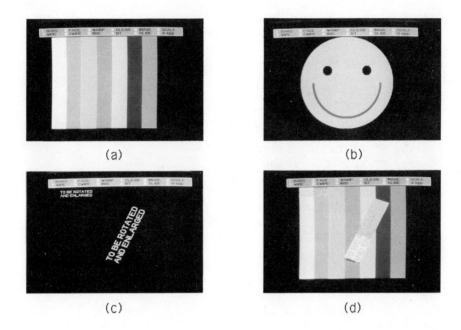

(a) (b)

(c) (d)

Figure 9.3   **Typical screens from demo.c. (a) color bars, (b) smiley face, (c) warp, (d) warp over color bars**

without clearing it first; thus, some functions will write on top of the results of other functions. The menu will include a choice for clearing the display area. When a function is completed, control will return to the menu automatically without requiring any user keystroke. The menu function will be set up so that it can be easily expanded to include more items which we will add in later chapters.

Photographs of all the screens generated by the examples in this chapter are shown by Figure 9.3. The C source code for the program begins in Listing 9.1. Note that there are some lines in the code which provide for features which we will add in later chapters. We'll point those things out as we come to them.

```
/*************** Start of Listing 9.1 ***************/

#include        <stdio.h>
#include        <conio.h>
#include        <string.h>
#include        "vdvi.h"

/*  Define the screen parameters  */
#define SC_WIDTH        512
#define SC_HEIGHT       480
#define MENU_X          10
#define MENU_Y          10
#define MENU_WIDTH      SC_WIDTH - 2 * MENU_X
#define MENU_HEIGHT     36
#define ENCODE_VER      1
#define MENU_PITCH      78

/*  Forward declaration of functions    */
I16     main( void);
I16     initialize( void );
I16     menu ( void );
void    clearBM( void );
void    do_bars( void );
void    do_face( void );
void    do_warp( void );
void    do_image( void );
I16     load_image( I16*, I16* );
void    do_image1( void );
void    do_scale( void );
void    do_wipe( void );
void    do_cwipe( void );
void    do_iris( void );
void    do_bt ( void );
void    do_slide( void );
void    do_9bpp( void );
I16     load_9bpp( hGrBm, char * );

/*  Declare global variables    */
I16     select,         /* stores the current menu selection */
        menu_drawn = 0, /* flag that menu is drawn           */
        text_setup = 0, /* flag to show that text is set up  */
        image  = 0;     /* indicates which image is loaded   */
```

```
hGrFont font_handle;    /*  Handle to the font used in menu   */
hGrBm   pBM,            /*  Bitmap which will be displayed     */
        pBMw;           /*  Another bitmap for work            */

char    *item[] = {     /*  List of text for menu items   */
        "1","BARS",         "2","FACE",         "3","WARP",
        "4","CLEAR",        "5","IMAGE",        "6","SCALE",
        "7","WIPE",         "8","CWIPE",        "9","IRIS",
        "A","BT",           "B","SLIDE",        "C","9-bpp",
        " ",
        } ;

GrPix   colors[8];  /*  Stores the color bar colors  */

/**************  End of Listing 9.1 ***************/
```

The first thing in any C program is a list of header files to be included. In addition to the standard C header files, there is a special header file **vdvi.h** for DVI which define all the constants, variables, structures, and functions used by DVI Technology. For the standard C functions we use, we will also need to include *stdio.h*, *string.h*, and *conio.h* from the Microsoft C include files.

Then, we define a number of constants which set up the screen size and the menu bar size. And, to please the compiler, we make forward declarations for all the functions which are used in the program. This keeps the compiler happy when it finds a reference to a function before that function has been defined. (This also declares some functions which will be covered in later chapters.)

The DVI headers provide type definitions for all of the special variables used by the system. In addition, they also redefine the standard C types such as **int (I16), long (I32), unsigned (U16)**, etc. This is a feature which will make it easier in the future to port DVI code to platforms where the default C types may have different byte sizes. For example, on some systems **int** defines a 32-bit integer, where on the PC platforms it is 16-bits. Using the DVI special types, you do not have to worry about **I16** ever meaning anything except a 16-bit integer.

Next, we declare the global variables needed for this program, including the names for two bitmap handles and one font handle. The text for the menu items is put in a character array **item[ ]**, which will make it easy to add items or change the menu in the future. We also declare an array **colors[ ]**, which will be used to store the colors of the color bar pattern in 16-bpp format. The other variable definitions are associated with management of the menu setup and the text setup.

Listing 9.2 is the function **main( )**, which of course is the top-level program.

```
/*************** Start of Listing 9.2 ***************/
I16     main()
{
I16     x;

DviBegin();

/* Perform initialization, quit if error      */
if ((x = initialize() ) < 0) {
      RtxPrintf("\nError #%d\n",x);
       goto terminate;
        }

/* Put a prompt on the standard output display screen   */
RtxPrintf("\n\nType a number, or ESC to quit\n");

/* Main program loop - exit if 0 returned by menu()     */
while (TRUE) {
        select = menu();
        if (select == 0) break;
        switch (select) {
        case '1':
                do_bars();
                break;
        case '2':
                do_face();
                break;
        case '3':
                do_warp();
                break;
        case '4':
                clearBM();
                menu_drawn = 0;
                break;
        case '5':
```

```
                        do_image();
                        break;
                case '6':
                        do_scale();
                        break;
                case '7':
                        do_wipe();
                        break;
                case '8':
                        do_cwipe();
                        break;
                case '9':
                        do_iris();
                        break;
                case 'A':
                case 'a':
                        do_bt();
                        break;
                case 'B':
                case 'b':
                        do_slide();
                        break;
                case 'C':
                case 'c':
                        do_9bpp();
                        break;
                default:
                        break;
                        }
        }

terminate:
DviEnd();
return(0);
}       /*  end of main()         */
/**************** End of Listing 9.2   ***************/
```

The structure is very simple: all DVI programs start with a call to **DviBegin( )**, which activates all DVI subsystems — this must be called first before any other DVI functions are used. Then the **initialize( )**

function is called — if that function returns a negative value, it means that there was an error during initialization and the program terminates. (Different errors during **initialize( )** will return different negative numbers—by interpreting the return number, the programmer can tell where something went wrong.) The function **RtxPrintf( )** is used to display an error message if necessary; this function is a special version of the standard C function **printf( )** which has been modified to work properly in the Rtx multi-tasking environment. (Several other standard C functions also have to be modified this way.)

If we get through **initialize( )**, we enter an endless loop: **while(TRUE)**. (**TRUE** is a constant defined in the DVI header as 1.) The first function in the loop is **menu( )**, which draws the menu and does not return until a selection has been made. The key character of the selection is contained in the global variable **select**. If **select** is zero, it means that the user selected Q, q, or ESC, and the program is terminated by breaking out of the **while(TRUE)** loop. Otherwise, the selected demonstration is executed by means of the **switch(select)** function after which control returns to the **menu( )** function for another choice by the user. The **switch( )** list includes selections 1–4, which will be explained in this chapter, and selections 5–B, which will be covered in Chapter 10. If the program is to be terminated (indicated by leaving the **while(TRUE)** loop, or by going to the label **terminate**:), we must call **DviEnd( )** before ending **main( )** to insure that the DVI functions are properly closed out. However, this does not disturb whatever is being displayed on the DVI monitor when we terminate the demo program.

Initializing

Initialization of a DVI program for graphics requires at least four steps:

1. Set up the DVI environment by calling **DviBegin( )**.
2. Allocate at least one bitmap structure by calling **GrBmAlloc( )**. (In this example we will set up two bitmaps.)
3. Initialize bitmap values by calling **GrBmSetPacked( )** or **GrBmSetPlanar( )**, and explicitly set any attribute values that are different from the defaults by using **GrBmSet( )**.
4. Set up the 82750 DA to the desired screen format by calling **GrBmDisplay( )**.

Listing 9.3 shows the **initialize( )** function that performs these tasks.

```
/************** Start of Listing 9.3 **************/
I16     initialize()
{
I16     x;

/*  Turn off the video display  */
x = GrBmDisplay(NULL);
if (x < 0) return(x);

/*  Allocate bitmap data structures  */
x = GrBmAlloc(&pBM);
if (x < 0) return(x);
x = GrBmAlloc(&pBMw);
if (x < 0) return(x);

/*  Describe the screen bitmap format:                */
/*    at address 0x10000 in VRAM,512x480 resolution, 16 bpp  */
x = GrBmSetPacked(pBM, GR_BM_16, 0x10000L, SC_WIDTH, SC_HEIGHT,
                  SC_WIDTH);
if (x < 0) return(x);

/*  Describe the work bitmap format:                  */
/*    at address 0x90000 in VRAM, 512x480 resolution, 16 bpp  */
x = GrBmSetPacked(pBMw, GR_BM_16, 0x90000L, SC_WIDTH, SC_HEIGHT,
                  SC_WIDTH);
if (x < 0) return(x);

/*  Fill the colors[] array with 16-bpp values       */
/*  (white,yellow,cyan,green,magenta,red,blue,black   */
GrPixFromColor(pBM, GR_COLOR_RGB, 255, 255, 255, &colors[0]);
GrPixFromColor(pBM, GR_COLOR_RGB, 255, 255,   0, &colors[1]);
GrPixFromColor(pBM, GR_COLOR_RGB,   0, 255, 255, &colors[2]);
GrPixFromColor(pBM, GR_COLOR_RGB,   0, 255,   0, &colors[3]);
GrPixFromColor(pBM, GR_COLOR_RGB, 255,   0, 255, &colors[4]);
GrPixFromColor(pBM, GR_COLOR_RGB, 255,   0,   0, &colors[5]);
GrPixFromColor(pBM, GR_COLOR_RGB,   0,   0, 255, &colors[6]);
GrPixFromColor(pBM, GR_COLOR_RGB,   0,   0,   0, &colors[7]);
```

```
/* Clear the bitmap to black  */
clearBM();

/* Display the bitmap */
x = GrBmDisplay(pBM);
if (x < 0) return(x);

return(0);
}
/************** End of Listing 9.3  ****************/
```

All the DVI functions return a status code value; if that value is zero or positive, the function was successful; if it is negative, there was an error. In **initialize( )**, all the DVI functions which could return runtime errors are tested for error and if there is an error, **initialize( )** will terminate, returning the error value to **main( )**. The negative return value is a code which gives some information about where and possibly why the error occurred; these codes can be interpreted (by a programmer) by using the DVI utility program **verrpr** which prints out the code explanation. In **initialize( )**, the variable **x** is used to hold the function return value; if **x** is negative, **initialize( )** returns immediately with the value of **x**. In the main program, the same approach is used to test the return value of **initialize( )** and abort if it is negative. The main program also prints out the error code value before quitting.

When our program starts up, we do not know what might be being displayed on the DVI screen. In order not to have spurious things displayed while we are initializing, it is usually desirable to blank the screen immediately. This can be done by calling **GrBmDisplay(NULL)**, which will blank the screen. (**NULL** is a constant which is defined as **0L** in the DVI header.) After all initialization is complete and we have filled the screen bitmap with the correct information, we can turn the display back on with another **GrBmDisplay( )** call.

The next step of initialization is to allocate system RAM for the bitmap structures we will be using. For this program, we will use two bitmaps: one to display, and another for a workspace that is not normally displayed. These bitmaps will remain for the duration of the program. In addition, some functions in the program will allocate temporary bitmap structures which get freed when the function ends. Two calls to

GrBmAlloc( ) get us the memory space in system RAM. (This has nothing to do with allocation of VRAM, which is not necessary.)

There are two functions for initialization of the values in bitmap structures — **GrBmSetPacked( )** and **GrBmSetPlanar( )**. These functions give the programmer control over the major parameters of bitmaps — other parameters are defaulted. If you want to change some of the defaults, there is a function **GrBmSet( )** which gives access to all bitmap data fields. *The Packed function sets up bitmaps where all the* components of each pixel value are located together in one contiguous group of bits — this is the form of 16-bit bitmaps. The **Planar** function sets up bitmaps where the three components of the pixel information (Y, V, U) are separated into three sub-bitmaps — this format is called *the 9-bit format* and it used by DVI Technology to achieve a high degree of compression; it will be explained in Chapter 11.

Initialization of bitmap structure **pBM** is for a screen format of 512×480 resolution with 16 bpp located at VRAM address 0x10000. This information is contained in the parameters for **GrBmSetPacked( )**. As explained above, this function allows the programmer to set the most common parameters of the bitmap and places default values in the other parts of the structure. For example, bitmaps can be located in VRAM, RAM, or EMS RAM, but **GrBmSetPacked( )** does not control that parameter. You have to make a separate call to **GrBmSet( )** to change the GrBmMemType field to anything other than VRAM. The second parameter tells the format of the bitmap — in this case 16 bpp. The third parameter is the address in VRAM for the bitmap; and the rest of the parameters define x-resolution, y-resolution, and pitch. A second call to **GrBmSetPacked( )** is made to initialize the working bitmap descriptor **pBMw**, whose bitmap is at VRAM address 0x90000. Note that this bitmap will go a little beyond the first megabyte of VRAM, meaning that this program requires at least 2 megabytes of VRAM. If we wanted to stay within 1 megabyte, we would have to reduce the size of the working bitmap, while leaving the top two pages of VRAM available for DVI system use. That would reduce our ability to prepare things off-screen before displaying them.

The next step is to initialize the array **colors[ ]** with the colors of the color bar pattern: white, yellow, cyan, green, magenta, red, blue, black. To get the color values, we use the function **GrPixFromColor( )**, which takes a bitmap pointer, a constant which tells what color component format will be used; then red, green, and blue values; and finally a pointer to the color variable **colors[n]** where the result is to be placed.

Because the bitmap pointer tells **GrPixFromColor( )** what the bpp format is, we do not need to worry about figuring out how to calculate the pixel value — this function does it. This will also work for more complicated pixel formats such as the 9 bpp video format. Because color pixel values are not the same for all formats, it is convenient in applications which use several different bitmap formats, to always use **GrPixFromColor( )** to get color values as they are used, rather than doing it ahead of time as we are doing here.

Another function is added to the initialization to clear the bitmap to black before we display anything. The function **clearBM( )** clears the bitmap **pBM** to black. It will be explained below. At this point, nothing is being displayed on the color monitor. In order to get the bitmap **pBM** displayed, we call the function **GrBmDisplay(pBM)**. This sets up the 82750 DA to the proper resolution and bpp values as specified by the bitmap structure, points the DA at the proper bitmap address, and turns on the display.

Clearing a bitmap to black

Listing 9.4 shows the source code for the function **clearBM( )** which fills a bitmap with black. This is our first encounter with one of the graphics primitive drawing functions.

```
/**************** Start of Listing 9.4 ********************/
/*  A function which clears the bitmap pBM to black       */
/*  It also waits until the DA has finished before returning */
void clearBM()
{
I16    x,y;

    /*  Set up the drawing attributes    */
    GrBmSet(pBM, GrBmDrawOutline, FALSE);
    GrBmSet(pBM, GrBmDrawColor, colors[7]);

    /*  Get the bitmap size  */
    GrBmGet(pBM, GrBmXLen, &x);
    GrBmGet(pBM, GrBmYLen, &y);

    /*  Perform the clear itself    */
    GrRect(pBM, 0, 0, x, y);
```

```
/*  Wait for the clear to be completed  */
      GrWaitDone( 100L );
}
/************** End of Listing 9.4    ********************/
```

Before we look at Listing 9.4, we need to discuss a problem which sometimes arises because of the way the runtime software sends commands to the 82750 PA. Since the PA is a co-processor which operates independently in parallel with the 80286 or 80386 CPU, we usually do not want to make the CPU wait for the PA to finish a task (or vice versa). Otherwise, there really would not be a benefit from the parallel architecture. Therefore, when we make a graphics call, the CPU figures out the correct commands for the 82750 PA and places them on a *display list* in VRAM. The graphics call then immediately returns, so that the CPU can go on to something else. Part of the PA's program RAM always has some microcode which reads the display list, and when it sees commands appear, it immediately begins executing them. In this way, the CPU can stack commands up for the PA and then begin another task while the PA independently carries out its own workload. In the case of **clearBM( )**, however, we want to be sure the bitmap is fully blanked out before we begin displaying it. Therefore we will deliberately make the CPU wait before **clearBM( )** is allowed to return.

To fill a bitmap (or any part of a bitmap — even just one pixel) with a color, we must first place the color value into the GrBmDrawColor field of the destination bitmap's structure using **GrBmSet( )**. The DVI system will pass the GrBmDrawColor to the drawing function itself. To clear the screen bitmap, we will use the **GrRect( )** function. This function must be told what kind of drawing to do — in this case, we want the rectangle to be *filled*, which is done by setting the GrBmDrawOutline field of the bitmap structure to FALSE using **GrBmSet( )**. Then we make the call to **GrRect( )**, passing the bitmap handle, and the coordinates for the entire bitmap. However, it is good practice to retrieve the size of the bitmap from the structure itself, so we added two calls to **GrBmGet( )** to do that.

The speed of **GrRect( )** with 16-bit pixels is about 6,000,000 pixels per second. Since our 512x480 bitmap contains 245,760 pixels, clearing that bitmap will take about 1/25 second. That is more than two vertical scans — in some cases this might cause a flash if clearing is done at the same time a bitmap is being displayed.

The last step of the **clearBM( )** function is to wait for the 82750 PA to finish. The DVI call **GrWaitDone( )** does just that. The argument of this function is the number of milliseconds (maximum) to wait. In this case, 100 milliseconds should be plenty.

The menu( ) function

The function **menu( )** is shown by Listing 9.5. Using the DVI text functions, **menu( )** displays a menu bar made from the list of items in the array **items[ ]**, and then it waits for a keystroke from the user.

```
/*************    Start of Listing 9.5    ******************/
I16      menu ()
{
I16      sel, x, y, count;
GrPix    color;


/*  Load the font, if it has not been done yet  */
if (text_setup == 0) {
    x = GrFontOpen ("sans.112", NULL, &font_handle);
    if (x >= 0) {
        x = GrFontLoad (font_handle);
        if (x >= 0) {
            GrBmSet (pBM, GrBmTextFont, font_handle);
            GrBmSet (pBMw, GrBmTextFont, font_handle);
            }
        }
    text_setup = 1;
    }


/*  Draw the menu if that has not already been done  */
if (menu_drawn == 0) {
    /*  Draw a light blue background rectangle for the menu  */
    GrPixFromColor (pBM, GR_COLOR_RGB, 70, 85, 235, &color);
    GrBmSet (pBM, GrBmDrawColor, color);
    GrBmSet (pBM, GrBmDrawOutline, FALSE);
    GrRect (pBM, MENU_X, MENU_Y, MENU_WIDTH, MENU_HEIGHT);
    /*  Draw a 2-pixel wide yellow box around the menu  */
    GrBmSet (pBM, GrBmDrawColor, colors[1]);  /*  yellow  */
    GrBmSet (pBM, GrBmDrawOutline, TRUE);
```

```
GrRect(pBM, MENU_X, MENU_Y, MENU_WIDTH, MENU_HEIGHT);
GrRect(pBM, MENU_X + 1, MENU_Y + 1, MENU_WIDTH - 2,
        MENU_HEIGHT - 2);

/*  Set up the text to be transparent  */
GrBmSet(pBM, GrBmTextBg, FALSE);

/*  Set up location to draw text from array: item[]   */
x = MENU_X + 10;
y = MENU_Y + 4;

count = 0;
/*  Draw items every 78 pixels , until a null item  */
while( (count < 24) && (*item[count] != (char) ' ') ) {
    if (count == 12) {
            y += 15;
            x = MENU_X + 10;
            }

    /*  Draw the item number or letter in white  */
    GrBmSet(pBM, GrBmDrawColor, colors[0]);   /*  white   */
    GrText(pBM, x, y, item[count]);
    count++;
    /*  Draw the item name in black  */
    GrBmSet(pBM, GrBmDrawColor, colors[7]);   /*  black   */
    GrText(pBM, x + 16, y, item[count]);
    x += MENU_PITCH;
    count++;
    }
menu_drawn = 1;
}

/*  Now wait for a keystroke   */
sel = getch();
/*  return 0 to quit if ESC or Q or q is hit  */
if ( (sel == 27) | (sel == 81) | (sel == 113) )   return(0);
return(sel);
}
/************** End of Listing 9.5     ******************/
```

Use of DVI Text Functions

The menu is drawn with DVI text functions, so we will explain them first. The text functions make use of font files which contain one bit-per-pixel bitmaps for each character of the font, with appropriate indexing so that the drawing function can find each character. Proportional-width characters are supported but bitmap fonts are not suitable for scaling to different sizes, so there has to be a separate font file for each character point size which will be used. Fonts in use are kept in VRAM, and the 82750 PA is programmed with microcode for the task of copying from the font array to the target bitmap with appropriate color or background modifications. Two functions are available for management of font files: one for identifying a font file, and the other for loading the font file into VRAM for use. There are two primitive functions for text drawing: one for drawing a text string to the screen, and another for checking how many pixels will be the length of a proposed string (without drawing it). However, using these two text primitives, more sophisticated functions can be written to provide formatting capabilities like the C function **printf( )**, or even to handle word wrap and other word processor types of formatting.

The first step in the **menu( )** function is to test the global variable **text_setup** to see whether text has been initialized. If it is zero, then the text initialization is done and **text_setup** is set to 1. Text initialization consists of three steps:

1. Make a call to **GrFontOpen( )** with the font name to get a handle to that font. This routine simply verifies that the font exists on hard disk and sets up the handle to refer to it later. If the font is not found, an error message is displayed, and the routine quits.

2. Then call **GrFont Load( )** to actually move the font into VRAM. This routine loads the font into a VRAM work space that is set up by the DVI runtime software when an application begins. This call is optional because subsequent calls to the text drawing routines will take care of loading the font into VRAM if it is not already there. However, this may lead to disk accesses occurring while drawing text, and that will slow things down. It is more manageable to make the call explicitly during initialization. The DVI runtime software supports a number of fonts simultaneously in VRAM, and (if you let it) it will automatically handle

management of them once the **GrFontOpen( )** call has been made for each one. Again, if **GrFontLoad( )** fails (indicated by its returning a negative value), an error message is printed and the routine quits.

3. The last step is to place the appropriate font handle into the GrBmTextFont field of the bitmap structure that will be used by the text drawing routines. Once all the fonts needed by an application have been loaded into the VRAM font buffer by steps (1) and (2), this step is the only one needed for further changing of fonts. (If the total space needed by all the fonts in an application becomes too large to fit in the VRAM buffer, the software will also correctly handle moving fonts from hard disk as needed.)

Once text initialization is done, the variable **text_setup** is set to 1, so that initialization will not be unnecessarily repeated on subsequent calls to **menu( )**. Note that the text initialization could have been been put into the subroutine **initialize( )**— we did not do that so that all the text functions would be together in one place in this example program.

Now we are ready to actually draw the menu. The variable **menu_drawn** indicates whether the menu has already been drawn. The first step is to put in a light blue background for the menu, using **GrRect( )** with the Filled attribute set. Because the light blue color is different from the blue of the color bar pattern (which is available in the **array colors[ ]**), we need to call **GrPixFromColor( )** to get the pixel value for light blue. Then a yellow rectangle two pixels wide is drawn around the background to provide a border, using **GrRect( )** with GrBmDrawOutline set to TRUE.

To draw the menu text, we first set up to draw text with a transparent background (GrBmTextBg=FALSE). The menu text is contained in the global array **item[ ]**; each menu entry uses two items in **item[ ]**: the first is a single character to identify the keystroke needed to select the entry on the menu, and the second item is the name of the entry. A **while** loop is used to draw menu entries until a blank is found in **item[ ]** or until 24 items are drawn. The **GrText( )** function actually draws the text. **GrText( )** is passed the bitmap structure to use, the x and y pixel locations within the bitmap to begin drawing, and a character pointer to the text string. DrawColor is set to white for the identifying character and to black to draw the entry name. On each pass through the loop, x is incremented by MENU_PITCH (78) pixels to change the text starting location to the next item. Since the menu box width is 492 pixels (see the #define for MENU_WIDTH in Listing 9.1), we can only get six menu items on one line using the 78-pixel menu pitch; therefore, if the count variable goes beyond 12, we move the y value for drawing text down by one line.

The fixed menu pitch of 78 pixels means that all menu text items should be less than 78 pixels in length, or the items will overlap on the screen, because **GrText( )** does not check anything about what you tell it to draw. However, there is another text routine which can be used to check how large a piece of text will be before it is drawn to the destination bitmap. This is **GrTextGetExtent( )** — it is passed the bitmap handle, a pointer to the text string, and a pointer to an I16 variable for the pixel length of the text if it were to be drawn.

If your application requires critical positioning of text, then the **GrTextGetExtent( )** call should be made before any text call; and the pixel length should be checked against any special requirements of the application. Using **GrTextGetExtent( )**, several programmers have written functions to do word wrap on text within a window and to center text or right justify it.

Having now drawn the complete menu on the screen, the rest of **menu( )** is concerned with capturing user response from the keyboard. Since we are interested only in single keystrokes, the standard C function **getch( )** (which waits for a keystroke) is appropriate. The character is tested for the ESC key (code 27) or the letter Q — if found, **menu( )** returns zero, which tells the main program to quit. Otherwise, **menu( )** returns the character for the key pressed.

Now we can look at one of the actual demo functions. Listing 9.6 shows the code for the **do_bars( )** function which does the color bar demo. We will draw an eight-bar pattern in the standard sequence shown in Figure 2.10. We will divide the screen horizontally and vertically into tenths, leaving one tenth all the way around for a border.

```
/************** Start of Listing 9.6  ******************/
/* Draw a color bar pattern   */
void do_bars ()
{
I16    x, y, xinc, ht,i;

/* Set up dimensions for bar pattern:
 * Divide both width and height into 10 parts,
```

```
      * bar pattern will use center 8 spaces in each direction.  */

      / Get width and height of bitmap */
      GrBmGet (pBM,   GrBmXLen,   &xinc);
      x = xinc /= 10;
      GrBmGet (pBM,   GrBmYLen,   &ht);
      ht = (ht * 8) / 10;
      y = ht / 8;

      /* Set up attribute for filled drawing     */
      GrBmSet (pBM,   GrBmDrawOutline,   FALSE);

      /* Draw the bars in left-to-right order using colors[] array */
      for (i = 0; i < 8; i++) {
        GrBmSet (pBM, GrBmDrawColor, colors[i]);
        GrRect (pBM, x, y, xinc, ht);
        x += xinc;
        }

      /* Draw a two pixel wide red border around the bars  */
      GrBmSet (pBM,   GrBmDrawOutline,   TRUE);
      GrBmSet (pBM, GrBmDrawColor, colors[5]);  /* red */
      GrRect (pBM, xinc, y, xinc * 8, ht);
      GrRect (pBM, xinc + 1, y + 1, (xinc * 8) - 2, ht - 2);


      }
      /************** End of Listing 9.6 ***************/
```

Since we already have the pixel values for the colors of the color bar pattern in the array **colors[ ]** (see Listing 9.1), drawing of the pattern is accomplished by one simple **for** loop. The bitmap field GrBmDrawOutline is set FALSE (outside of the loop) to draw filled rectangles; and then the loop places a color into the GrBmDrawColor field from **colors[i]**, calls **GrRect( )**, and increments the **x** drawing location by the amount **xinc** to get ready for the next bar. The rest of **do_bars( )** draws a two-pixel red rectangle around the bar pattern to make it look prettier. **do_bars( )** executes so rapidly that the bar pattern appears almost instantaneously even though it is being drawn directly to the display—the sequential drawing of the bars from left to right is so fast that it can hardly be noticed.

Drawing of Ellipses

The **do_face(**.**)** function, which draws a smiley face using the **GrEllipse( )** function, is shown in Listing 9.7. First, we define some variables for the sizes of the face and the smile. The values have taken into account the x and y resolutions and the aspect ratio in order to define approximate circles.

Then we can draw the outline of the face, filled with yellow. This is done by setting DrawColor and Filled in our attribute descriptor, and then passing bitmap, attribute, and a rectangle specification (**x, y, xsize, ysize**) to **GrEllipse( )**. **GrEllipse( )** draws an ellipse which just fits inside of the specified rectangle. The x and y locations of the rectangle for the face are calculated in the parameters so that the face will be centered on the screen, both horizontally and vertically.

```
/*************** Start of Listing 9.7 *************/
/* Draw a smiley face */
void do_face()
{
I16    facewidth = 310, faceheight = 380,
       smilewidth = 230, smileheight = 300;

/* Draw the yellow circle for the face */
GrBmSet(pBM, GrBmDrawColor, colors[1]);      /* yellow  */
GrBmSet(pBM, GrBmDrawOutline, FALSE);
GrEllipse(pBM, (SC_WIDTH - facewidth)/2,
         (SC_HEIGHT - faceheight)/2, facewidth, faceheight);

/* Doing the smile...                              */
/* Set up drawing attributes for off-screen bitmap */
GrBmSet(pBMw, GrBmDrawOutline, FALSE);
GrBmSet(pBMw, GrBmDrawColor, colors[1]);
/* Clear the off-screen bitmap to yellow */
GrRect(pBMw, 0, 0, SC_WIDTH, 384);
/* Now draw a red circle the size of the smile    */
GrBmSet(pBMw, GrBmDrawColor, colors[5]);    /* red */
GrEllipse(pBMw, 0, 0, smilewidth, smileheight);
```

```
/*  and draw a yellow circle inside to leave a ring of red  */
GrBmSet(pBMw, GrBmDrawColor, colors[1]);    /*  yellow  */
GrEllipse(pBMw, 8, 10, smilewidth - 15, smileheight - 20);
/*  now we will transparently copy the bottom half of the ring
 *  to the screen to make the smile.                        */
GrBmSet(pBM, GrBmCopyKeyColor, colors[1]);    /* yellow */
GrBmSet(pBM, GrBmCopyKey, TRUE);
GrCopy(pBM, (SC_WIDTH - smilewidth)/2,
faceheight/2 + (SC_HEIGHT - faceheight)/2, smilewidth,
faceheight/2, pBMw, 0, smileheight/2);
GrBmSet(pBM, GrBmCopyKey, FALSE);

/*  and draw the eyes...        */
GrBmSet(pBM, GrBmDrawColor, colors[7]);    /* black */
GrEllipse(pBM, SC_WIDTH/2 - 80, SC_HEIGHT/2 - 70, 30, 40);
GrEllipse(pBM, SC_WIDTH/2 + 50, SC_HEIGHT/2 - 70, 30, 40);
}
/************** End of Listing 9.7 ***************/
```

Now, we have to get a little tricky to draw the smile. Since
GrEllipse( ) always draws complete ellipses, we need to go to the off-
screen bitmap **pBMw** and create the smile as a full circle; then we can
copy the part of the circle we want for the smile transparently over the
face outline which is already on the screen. To do this, we first call
GrRect( ) to fill the work bitmap with yellow. Then, with **GrEllipse( )**,
we draw a circle of red (the smile color), which has the outside diameter
of the smile. A second call to **GrEllipse( )** draws a yellow circle inside
of the red circle, leaving a ring of red.

The last part of the smile process is to do a transparent copy of the
bottom half of the red ring onto the yellow face which is on the screen.
This is done using **GrCopy( )**, with the GrBmCopyKey field set to TRUE
(transparent), and the GrBmCopyKeyColor field set to yellow. This
means that **GrCopy( )** will not copy any yellow pixels from the source
bitmap. Thus we can copy a yellow rectangle containing the smile arc
without any of the yellow rectangle showing on the screen. Since this is
our first use of **GrCopy( )**, we should explain its parameter list, which
may be a little confusing. The destination bitmap handle, and a
destination rectangle specification are passed first. Then you pass the
source bitmap handle and the x,y location in the source bitmap to begin

copying. (Of course the source rectangle size is the same as the destination rectangle size, which has already been specified.)

To complete the smiley face, we do two more calls to **GrEllipse( )** with GrBmDrawColor set to black and GrBmDrawOutline set FALSE, to draw the eyes. Again, the DVI graphics functions are so fast that the assembly of this graphic seems almost instantaneous. There is a brief delay noticeable between the drawing of the yellow outline and the rest of the face appearing — this is the time during which the smile is being drawn off-screen. That delay could be made invisible by drawing the smile off-screen before drawing the yellow outline. This is a general technique for making graphics operations look fast — assemble as much of the graphic off-screen first and then copy it to the screen with one or more **GrCopy( )** calls.

Use of the Warp Function

Another part of **demo.c** is shown in Listing 9.8; it rotates text on the screen using the **GrPolyWarp( )** function.

```
/*************** Start of Listing 9.8 *************/
/*  Write a line of text on the screen,
 *  use WARP to rotate it 90 degrees */
void do_warp()
{
hGrBm    pBMsrc;
GrPix    color;

    /* Set up arrays holding warp source and dest coordinates */
I16     xs[4], ys[4], xd[4], yd[4];
        xs[0] = 1;       ys[0] = 1;
        xs[1] = 1;       ys[1] = 31;
        xs[2] = 126;     ys[2] = 31;
        xs[3] = 126;     ys[3] = 1;

        xd[0] = 200;     yd[0] = 350;
        xd[1] = 250;     yd[1] = 370;
        xd[2] = 350;     yd[2] = 120;
        xd[3] = 300;     yd[3] = 100;
```

```
/* set up a 128-pitch bitmap for the Warp source */
GrBmAlloc(&pBMsrc);
GrBmSetPacked(pBMsrc, GR_BM_16, 0x90000L, 128, 128, 128);

/* Write two lines of text to the screen in a gray color */
GrPixFromColor(pBM, GR_COLOR_RGB, 127, 127, 127, &color);
GrBmSet(pBM, GrBmDrawColor, color);
GrText(pBM, 50, 60, "TO BE ROTATED");
GrText(pBM, 50, 75, "AND ENLARGED");

/* Copy those lines of text to the Warp source bitmap */
GrCopy(pBMsrc, 0, 1, 127, 32, pBM, 49, 58);

/* Do the warp to the center of the screen */
GrPolyWarp(pBM, xd, yd, 4, pBMsrc, xs, ys);

GrBmFree(pBMsrc);
}
/************** End of Listing 9.8 ***************/
```

GrWarp( ) is a mapping function which distorts one polygon onto another polygon with the same number of points.   **GrWarp( )** can operate between two polygons in the same bitmap or between two different bitmaps. (Most graphics functions work on any bitmap size or pitch as long as the pixel formats are the same. The **GrCopy( )** function can even copy automatically between bitmaps of different pitch or different resolution, but not with different bpp values.)

To demonstrate in this example the flexibility of the graphics functions to work with different size bitmaps, we will write text to the screen, then copy it to a Warp source bitmap which has a different size and pitch, and then Warp it back to the screen. We will use four-sided polygons: the source is a rectangle enclosing the text, and the destination is another rectangle which is rotated to a different angle.

Looking at the code in Listing 9.8, we first set up a temporary bitmap, **pBMsrc**, which will be used as the Warp source bitmap with a pitch of 128 pixels. (We will remove this from memory at the end of this function.) Then we declare and initialize four arrays for the pixel coordinates of the source and destination polygons.   The coordinate points of the two polygons are put in these arrays in counterclockwise order going around the outside of each polygon. Both source and destination polygons must have the same number of vertices. In many applications, the source and destination coordinates would be calculated according to some algorithm — for example, a 3-D perspective algorithm. However, for simplicity in this example, the source and destination values are hard-coded here.

Then we allocate memory and initialize the bitmap with calls to **GrBmAlloc( )** and **GrBmSetPacked( )**. Following this, we draw two lines of text on the screen beginning at coordinates 50,60, using transparent mode, which is obtained by setting GrBmTextBg to FALSE. The next step is to copy this text from the screen to the Warp source bitmap using **GrCopy( )**.

Finally we call **GrWarp( )**, passing destination bitmap handle, destination coordinate array pointers, number of points (4), source bitmap handle, and source coordinates array pointers. The function is then completed by calling **GrBmFree( )** to deallocate the memory used by the temporary bitmap descriptor.

Points Made by Software Example

The previous software example was included to demonstrate to programmers that DVI software is easy to use. The runtime software takes much of the detail of a powerful video and graphics system away from a programmer who does not want to worry about details. At the same time, all of the system capability and speed is still available. For example, the graphics fill speed using the runtime software routines is 6,000,000 16-bit pixels per second, which is close to the raw speed of the 82750 PA. This happens because the runtime software sets up the function and then hands the task over to PA microcode, which can then run at PA speed and has no further CPU software overhead.

It is difficult to convey the speed of DVI graphics without actually seeing execution of the simple functions described above. For these kinds of operations, the apparent speed of execution is largely determined by considerations other than the graphics drawing speeds. In particular, the first time that any graphics function is called, the DVI system software has to load that function's microcode from hard disk into VRAM. Therefore, there is a brief delay while a disk access occurs. However, for subsequent calls to the same function, there is no delay because the microcode is already in VRAM.

82750 PA Microcode Software

We have made many references to the microcode programs which run on the 82750 PA. Much of the flexibility of DVI Technology exists because of what can be done with PA microcode. The PA is a fast processor itself, running at 12.5 MIPS, but that speed is made even more powerful when you consider that via the microcode you can essentially reconfigure the PA for each instruction cycle! However, that same statement says that programming in microcode inherently requires an understanding of PA architecture, and it is not going to be as easy as programming in C.

There are tools available from Intel for programming microcode, and some users will undertake their own algorithms, because they really need them. This will require programmers who are prepared to deal with all the detail that microcode involves. The primary microcode programming tool is an *assembler* program which allows the writing of microcode in a mnemonic language. The assembler keeps track of most of the housekeeping, but it still takes an intimate knowledge about what the PA is doing for each instruction. Along with the assembler, there is a debugger program, which is available with the microcode development software. In addition to commands for exploring what is going on in the PA, the debugger has an extensive capability to examine and manipulate VRAM, which can also be useful for debugging C programs.

It is not out of the question to write a C compiler for the PA microcode. That possibility is being considered for inclusion in a future release of the DVI software. It would increase the number of people who could create custom microcode programs.

Once a custom microcode program has been produced, it is very easily integrated into a C program for use just like any of the microcode programs which are already in the runtime package.

DVI Audio Software

It should be evident from the hardware discussion of Chapter 8 and the earlier part of this chapter that DVI video and audio hardware and software are conceptually similar. The audio system contains its own powerful processor and its own dedicated memory just like the video system. The same similarity applies to the software. There is a driver and a driver interface module for each system.

The runtime software takes care of setting up the audio algorithm software in the TMS 320C10 audio processor. Audio is played by AVSS, and there is full flexibility for integration of the audio with video. There is a complete set of audio capture and conversion tools to allow creating and working with AVSS (.avs) audio-only files. Similarly, the audio/video editing tools allow the combination of audio and video and other material into one AVSS file. Audio has not been forgotten in the DVI software — it has all the flexibility and functionality needed to go along with the DVI video capabilities.

Summary

This chapter has given an overview of the DVI software structure and a first example of a DVI application program which uses some of the Graphics functions. The example should have convinced you that DVI programming is not very complicated and that the Graphics primitive functions are easy to understand and use; they are fast, and they are powerful. In later chapters, we will be expanding upon this sample program to include some of the other DVI capabilities beyond simple graphics.

10

Still Images

There are many applications where photographic-quality still-picture reproduction is a key requirement. For example, a personnel data base for security purposes ought to be able to display photographs of the people. Similarly, a real estate data base needs to show pictures of the houses that are for sale, along with all the numerical information about the property. Any presentation is enhanced by the inclusion of photographs at key points — we are used to this in print media, so it's only natural that we should want the same thing in our state-of-the-art computer.

Any digital video system has some degree of still-picture capability — since it requires no special processing to simply display still pictures as long as you are not in a hurry. All that it takes is to load image pixel data (in the right format) from disk into the memory that is being used for display. However, when the system has a powerful video processor such as the 82750 PA of DVI Technology, still picture reproduction can be substantially enhanced with respect to picture quality, display flexibility, storage capacity, and speed of retrieval. This is because we can

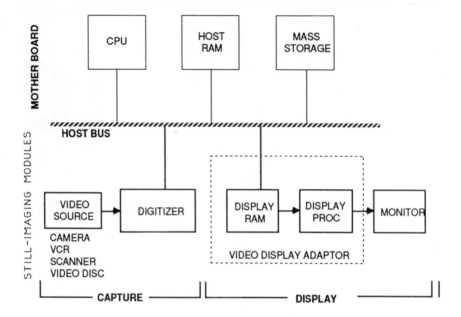

Figure 10.1   Still image hardware block diagram

use the video processor to compress and decompress images and to manipulate the images before they are displayed (or *as* they are being displayed). Of course the same processes could be programmed on the host CPU of the computer, but with host processor power in the range of the 80286 or 80386, this is an order of magnitude or more too slow to be practical.

Figure 10.1 shows the typical hardware components of a digital video still-image system. The host computer is shown at the top of the diagram, and the still imaging hardware is shown below. In the case of a PS/2 or a PC AT host system, all the hardware below the bus line would be on one or more plug-in modules or peripherals. Most PCs have the video display adaptor as a plug-in card, although the IBM PS/2 systems have a VGA display adaptor built in to the mother board. It is still possible to add another display adaptor as a plug-in — this is exactly what happens with the ActionMedia 750 Delivery Board in a PS/2. (Thus a PS/2 automatically has DVI and VGA both when the ActionMedia card is used.)

Video Digitizing

The digitizer (A/D converter) is an expensive item, and therefore it is often an optional feature — usually in the form of a piggyback board on the main video display module, or as a separate unit. The video source for digitizing can be from a video camera, from a tape recorder, from a video disc, or any other signal of the proper format. Digitizers are available for either RGB or composite formats.

Another type of input device is the *scanner*, a device which converts a hard copy or transparency directly to digital video. Scanners usually have a solid-state *line-scan pickup device* which electronically scans horizontally at extremely high resolution (300 dots per inch is typical). Vertical scan is provided by mechanically moving the line-scan device relative to the original. The line-scan device delivers an analog output, but A/D conversion is usually included inside the scanner. Therefore, a scanner combines the functions of video pickup and digitizing in one box. Image scanning is somewhat slow — processing an 8x10 original may take from 30 seconds to several minutes, but the results from the available devices are excellent. Color image scanners, although they can be expensive ($5,000–$10,000 or more), will provide the best quality digital output from hard copy or transparency originals.

There are several important considerations for selection of original materials and techniques for still-image digitizing. For highest quality of the digitized output, the input material must be of the highest quality. This is most important for high-resolution digital formats (512x480 or above) or if the digital image is going to be compressed. This quality consideration means that the camera quality should be broadcast-level, and RGB is preferable to NTSC or PAL composite. (The ActionMedia Capture Board has an RGB input.) The signal-to-noise ratio must be high (46 or better) because still-image capture is also going to capture the noise of a single frame, and a stationary noise pattern is more visible than noise which moves randomly. Both the resolution and noise considerations rule out low-priced VCRs — digitizing from such a source produces poor results. When there is a choice, go for the highest quality video source equipment that you can afford.

There is a problem in digitizing video when the source is a film image (transparency or print). This is because of an inherent mismatch between the characteristics of the additive color reproduction of electronic video and the subtractive color reproduction of film systems. As explained in Chapter 2, high-quality video reproduction from film requires the use of a special video camera called a telecine camera. If a

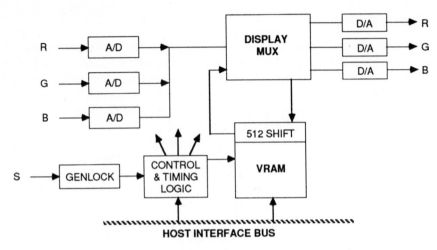

Figure 10.2　Targa board block diagram

telecine camera is available, it should be used as the source for digitizing from slides or film.

It is also possible to do some correction of the image from film in the digital domain after it is digitized. Some of the scanner devices have this capability in their software, and systems with video processors such as DVI Technology can be programmed to do gamma correction and color correction for the film-to-video interface problems.

Truevision Targa boards

There are a number of still-image video systems on the market in the form of PC plug-in display adaptor modules. A good example is the Targa series of boards produced by Truevision, Inc. These boards do not have a programmable processor like the 82750 PA, but they have been widely applied to still image applications. The Targa boards contain their own video RAM, and offer options for digitizing from RGB or composite inputs. They have a hard-wired display controller, with built-in features for video display and manipulation. Figure 10.2 is a simplified block diagram of a Targa board, which we will briefly describe.

Targa boards are available in 8-bpp, 16-bpp, 24-bpp, and 32-bpp versions, becoming progressively more expensive because of the need for more RAM at higher bpp and because the higher-priced boards have more features. The Targa boards use VRAM chips for their display memory. The serial shift-register output from the VRAM chips is used to refresh the display just as in the DVI video board. The shift-register

(a)

(b)

Figure 11.4 Performance of DVI still image compression using vcomp program.
(a) 768x480 image at 414,720 bytes (9bpp), (b) 768x480 compressed to 92,160 bytes (2 bpp). (Original image copyright Impact,
photographs by Brad Wagner.)

Figure 13.7 DVI screen photograph of the 9-bpp demonstration mode of demo.c. (512x480 resolution, 9 bpp)

(a)

(b)

Figure 10.4 Results of DVI image demonstrations in demo.c. (a) selection 5 IMAGE, (b) selection 6 SCALE. (512x480 resolution, 16 bpp)

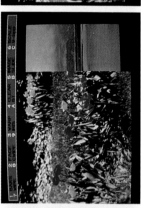

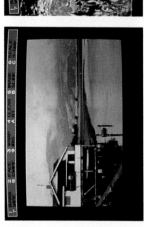

Figure 13.1   Sequence of DVI screen photograph taken during a wipe transition. (512x480 resolution, 16 bpp)

Figure 13.2   DVI screen photograph taken during
a cursor wipe tran-
sition. (512x480
resolution, 16 bpp)

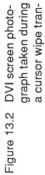

Figure 13.6   Sequence of DVI screen photographs taken during a slide-on transition.
(256x240 resolution, 16 bpp)

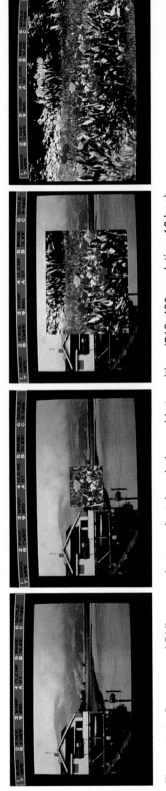

Figure 13.3 Sequence of DVI screen photographs taken during an iris transition. (512x480 resolution, 16 bpp)

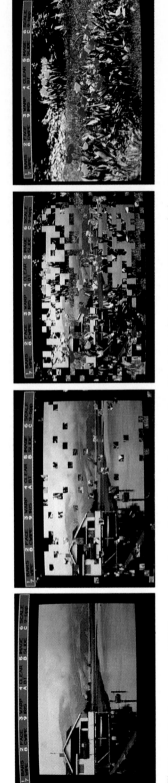

Figure 13.4 Sequence of DVI screen photographs taken during a random-block transition. (512x480 resolution, 16 bpp)

is loaded one line at a time during horizontal blanking and then refreshes the display monitor during active scan time.

Targa boards are designed for RGB input and output (the Targa-16 also has NTSC composite input/output), and there are three A/D converters on the input and three D/A converters at the output. In the Targa-16, which uses RGB 5:5:5 pixel format, the A/D converters are 5 bits; in the other models they are 8 bits. A display multiplexer (switch) allows display of either the input passed through from the A/Ds (live mode) or whatever is in the display RAM. The display RAM stores one full-resolution page (512x482 active pixels), which can be captured from the input or accessed from a storage medium via the host computer interface. There is also genlock circuitry so that the display may be synchronized with incoming video.

This relatively simple hardware architecture gives a lot of flexibility —video from computer storage or computer generation can be keyed into live video from the input. Live video can be captured as still frames. By modifying the way in which the display shift register operates, a zoom feature is implemented to enlarge the image by 2x, 4x, or 8x.

Truevision also supports the boards with a good developer's package of software which allows an application programmer to access all the features easily. Furthermore, there are now many application packages available for these boards to provide paint and drawing capability, as well as other more specialized capabilities which depend on display of realistic video stills. However, the Targa features are hard-wired, and new features can only be done by means of software on the host CPU, which is going to be slow. This point is well appreciated by Truevision, Inc. and they have come out with a high-end family of boards (called Vista) which include a Texas Instruments TMS 34010 programmable graphics processor chip, which promises to offer much more speed and flexibility.

DVI Technology for a Still-Image System

The DVI hardware contains all of the ingredients of a still image system as shown in Figure 10.1. With appropriate programming, DVI Technology can reproduce excellent still images, and it offers tremendous flexibility because it is software based. The following discussion covers the DVI software for still images. This section will go deeply into the software area, and nonprogrammer readers may skim over it and go on to the section **Points Made by the Software Example**.

DVI image files

In this chapter we are talking about still images, by which we mean stationary images which are stored in *image files*. An image file is a special kind of binary file which contains the complete pixel data for a single picture or a single screen. There are many formats for image files, depending on the system which creates them and on the particular resolution and bpp values which are used. Image files may be originated by digitizing single frames from a video camera source, by reading from hard copy or a photograph using a scanner, or simply by saving the data from a bitmap which may contain graphics, text, or other information. There are also special types of image files which store the data for compressed images. We will cover that in Chapter 11.

Most of the digitizing hardware described above produces a digital image in RAM which can be immediately displayed or stored to disk. Some systems also can process or modify the stored image to enhance some aspect or to change size, etc. However, the end result in all cases must be a file which is stored for later use. The simplest kind of image file is just a *RAM dump* of the information in the bitmap being displayed — the entire bitmap, byte-for-byte, is stored on disk in a binary file.

Even the simplest image file needs to have some header information if there is to be any flexibility about how that file is loaded and displayed. If the file contains a bitmap as described above, then we should include information in the file to describe the bitmap format, preferably right at the start where it will be easy to find. Of course a system could be designed for only one bitmap format and all that information would be assumed for every file — that is a no-flexibility approach. It is much better to at least include the x and y pixel resolution values in the file header; and if the display system has different bpp modes, the bpp value also ought to be in the image file header. Further flexibility is offered by adding even more information to image file headers.

The DVI software uses image files for storing bitmap data as described above as well as for storing compressed images which are not bitmaps. Many formats are available, indicated by information contained in the file header (which will be described below), and also by the filename extension. The following is a list of DVI image file formats with their filename extensions:

Component formats:
.imr   Red component, 8 bpp
.img   Green component, 8 bpp

.imb Blue component, 8 bpp

.imy Y luminance component, 8 bpp
.imv V color component, 8 bpp
.imu u color component, 8 bpp

.imm Monochrome gray-scale, 8 bpp

.ima Alpha channel data, 8 bpp

Device-specific formats: (where the program accessing these files must have built-in knowledge of the exact pixel format)
.i8 8-bit device-specific pixels
.i16 16-bit device-specific pixels (for example, the DVI 16-bpp format, which is device-specific)

Color map formats:
.imc Color map data, optionally including 8-bit index pixel values

Compressed image formats:
.cmy
.cmv
.cmu Compressed color component data (last character of extension indicates which one)
.c16 Compressed 16-bit device-specific data

The header system for DVI image files is extremely flexible, containing:

- Complete information about the file's content, format, and version
- Provision for future enhancements to be added to the system or to the header structure while maintaining backward compatibility with older versions
- File management features

The DVI image file header concept is described in Figure 10.3. Image file headers should always be accessed through the **GrImHdrGet( )**

START →
OF FILE

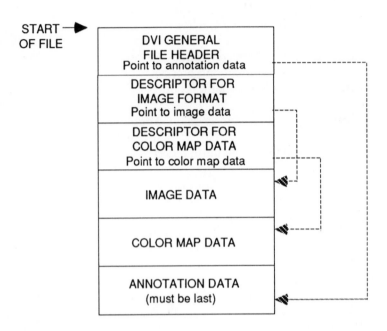

Figure 10.3   DVI image file structure

function or (in the case of the standard header) through pointers using the field names given in the header file **vstd.h**; therefore, this description only shows the logical structure of the header. This restriction on access methods is critical to maintaining the DVI approach to compatibility with future enhancements.

The first part of the image file is a general file header structure which is used for all DVI file types. There are four fields in the general header; one is called FileID, a four-byte character string which indicates the type of file — for image files FileID is "VIM". Another field is HdrSize, which defines the size (length) of the entire file header in bytes. Another field, HdrVersion, is a code number which identifies the version — this number is currently 5 for image files; previous versions (1–4) cover the stages of development of this format since 1983!

The fourth entry in the general file header part of the image file is a 32-bit pointer AnnOffset to the byte location of annotation text for this file. Normally any annotation will be added to the end of the file (after the image data), and it could include information such as the source of the original video, the type of camera used, original capture data, or anything else that the developer wants to include. Since the annotation is at the end of the file, additional information is easily appended if the file undergoes any modification. If annotation is not used, AnnOffset will be NULL. Note that all files in the DVI system — audio, video, fonts, microcode, etc. — not just images, use the standard header just described and therefore they all have the features described so far.

Now we come to the beginning of the image-file specific part of the header; the first part describes the image format. The fields include GrImHdrColor, which is a flag field encoded to define the color format of the file (RGB, YVU, monochrome, etc.); and GrImHdrPlane, the color planes included in the file (R, G, B, RGB, Y, V, U, YVU, etc). These parameters define the format of the file and should be consistent with the filename extension as described above.

Other fields are the GrImHdrXLen and GrImHdrYLen which specify the pixel size of the image, and the field GrImHdrPixelBits, which is the bpp value for the file. Another field is GrImHdrAlgVer; for encoded files, this tells the system software what software to use for decoding the file. This field is zero if the file is uncompressed pixel data.

Further we have the GrImHdrImageOff field, which is a byte pointer from the beginning of the file to the start of the actual video data in the file. If this pointer is NULL it means that the file contains no video data.

The remaining defined field of the image descriptor is the GrImHdrImageSize field, which is the byte size of the image data. For uncompressed images, of course this could be calculated from the XLen, YLen, and PixelBits fields; but that will not work for compressed video data files.

Three entries in the header are used when the file contains a color map, which could either be by itself or in addition to video data. GrImHdrClutCnt is a count of the color map entries in the file; GrImHdrClutBits is the number of bits per color map entry; and GrImHdrClutOff is the byte offset from the beginning of the file to the start of color map data.

This arrangement of the file header using offsets to the actual data locations allows the header to be modified in future DVI software versions; but as long as existing header items are not moved (add new items only at the end), files created by a new version will still be readable by older software — lacking the new features, of course. There also is a lot of flexibility for including other data along with the images, by means of the Annotation feature.

Image file format conversion
We have described the flexibility of image file formats in DVI Technology, and we should note that other image systems, such as the Targa boards, have different image file formats. All of this variety clearly points out the need for utility programs that do image file format conversion. Of course, we would have to accept that conversion of a high-quality format to a lower-quality format (such as converting 32 bpp to 16 bpp) is going to cause an irretrievable loss of data; however, file conversion is still extremely useful; in fact it is in a way a form of compression.

Intel provides an image file conversion utility for DVI Technology — the program **vimcvt**. This program can take files in any of the following formats as input and convert them to another format from the same list as output:

Code	Format	Files/Image
i16	DVI 16-bpp format	1
c16	DVI 16-bpp compressed	1
yvu	DVI 9-bit YVU format	3
yiq	DVI 9-bit YIQ format	3
c9	DVI 9-bit YVU comp'd	3
rgb	DVI 24-bit RGB format	3
avs	DVI AVSS format	1
t16	Targa 16-bpp format	1
t32	Targa 32-bpp format	1
pix	Lumena DVI format	1

(Lumena is a professional-quality paint program that has been ported to DVI Technology.) **Vimcvt** does the best that it can in the conversion; but as explained above, it cannot create data in a higher-quality format that was not contained in the input format. The resolution values for the image are unchanged by the conversion process. **Vimcvt** runs on the 82750 PA and conversion is quite fast. For example, converting a t32 file to c16 format at 512x480 resolution takes about 22 seconds on a 20-MHz 80386 machine if the source file is on hard disk.

Loading image files in DVI
Listing 10.1 shows a C function **do_image( )** which loads one of two image files **tul.i16** (picture of tulips) or **coast.i16** (picture of house on the coast) and then immediately displays it; this function is included in

demo.c as menu item number 5.

```
/*************** Start of Listing 10.1 *************/
/* Display one of two images, different each time  */
void    do_image()
{
I16     x,width,height;

/* Load the image to BMw and return its size  */
x = load_image(&width, &height);
if (x < 0) return;

/* Now copy the image to the screen, centered */
GrCopy(pBM, (SC_WIDTH - width)/2, (SC_HEIGHT - height)/2,
       width, height, pBMw, 0, 0);
}


/*----------------------------------------------------------*/
/* Load an .i16 image to BMw, return its size  */
I16    load_image(xsize, ysize)
I16     *xsize,*ysize;
{
I16     x;
char    *fileName0 = "tul.i16";
char    *fileName1 = "coast.i16";
char    *fileName;
hGrImHdr ImHdr;

/* select a different image to load  */
if (image == 0) fileName = fileName1;
else fileName = fileName0;
image ^= 1;

GrImHdrAlloc(&ImHdr);

/* Look for the file and get some data about it       */
x = GrImLoadHdr(fileName, NULL, pBMw, ImHdr);
if (x < 0) return(-1);
GrImHdrGet(ImHdr, GrImHdrXLen, xsize);
GrImHdrGet(ImHdr, GrImHdrYLen, ysize);
GrImHdrFree(ImHdr);
```

```
if (*ysize > (480 - 2 * (MENU_X + MENU_HEIGHT)))
    *ysize = 480 - 2 * (MENU_X + MENU_HEIGHT);

/*  Load the image into BMw    */
x = GrImLoad(fileName, NULL, pBMw, 0, 0, *xsize, *ysize);
if (x < 0) return(-2);


}
/************** End of Listing 10.1 **************/
```

The two images are digitized photographs which are stored on the hard disk in 16-bpp format image files. The extension .i16 on the filename indicates the 16-bpp format, which is the same format we used for bitmaps in the **demo.c** program described in Chapter 9. Therefore these images can be loaded directly into the bitmaps of **demo.c** without any conversion or processing.

The image **tul.i16** is 507x405 pixels — giving a 16-bpp image file of 410,730 bytes; the image **coast.i16** is smaller — 436x317 pixels or 276,484 bytes. The reason for the unusual pixel sizes of these images is that they were digitized at 512x480 resolution originally, but then they were cropped to include only the most interesting information. Loading uncompressed images in that size range from hard disk to VRAM takes 3 or 4 seconds. In order to provide a faster transition on the screen, we first load the file to the working bitmap **BMw**, which is not being displayed. Then we use **GrCopy( )** to bring the image to the right location on the screen. The transition then takes about 80 milliseconds, which is fast enough to appear instantaneous. Of course, in this example we have to wait the 3 or 4 seconds while the image loads off-screen before the transition occurs; in a real application we would usually arrange to load the image into VRAM during an idle period ahead of time so that the user would not need to wait when the image was selected. Using this same strategy, we could do a more elaborate transition, such as (for example) a wipe, a random block transition, or a mosaic transition; some of these *special effects* will be described in Chapter 13.

Looking now at the listing, the function begins with declarations of local variables to hold the image size and a working variable **x**. We then call the function **load_image( )**, which we will explain below. This function loads one of the two images to the working bitmap. We now have an image loaded from hard disk into our VRAM off-screen bitmap. The final step of displaying it is to copy it to the screen bitmap. This is done with a call to **GrCopy( )**. The positioning of the image in the destination bitmap (**pBM**) is calculated to center the image on the screen. Figure 10.4(a) shows what the resulting screen looks like (see color plates).

The function **load_image( )** is also in Listing 10.1. It is set up to load a different one of two images every time it is called, using the global variable **image** as a flag to keep track of which image was loaded last. The reason for making this a separate function is that we will use this function again later in different demonstrations of image operations. In this function we first set up local working variables, including the names of the two image files we will be alternating between, **fileName0** and **fileName1**. A character pointer variable **fileName** is used to point to the current filename. The first operation in the function is to set **fileName** to point to the right name, in accordance with the variable **image**, which can be either 0 or 1. Then **image** is toggled by an exclusive-or operation with 1.

We will get the image size from the image file header by loading the file header into memory and then using **GrImHdrGet( )**. We have to first allocate memory for the header by calling **GrImHdrAlloc( )**, and then the header can be loaded by calling **GrImLoadHdr( )**. Then we can use **GrImHdrGet( )** to extract the image size parameters **xsize** and **ysize**. At this point we can free the memory for the file header by calling **GrImHdrFree( )**.

All of the DVI routines which access image files use the following search strategy:

(1) First search for the file in any path specified with the filename.
(2) If there is no path specified, search the current directory.
(3) If the file is not found, search any paths specified in the DOS environment variable VIM.

Thus image files may either be located in the current directory, or they may be placed in a central image file directory (for example C:\IM) which is identified by the DOS environment variable VIM (defined, for example, by the DOS command: SET VIM=c:\im). This ability to redirect the DVI routines to different directories is convenient in certain situations.

We test the return value from **GrImLoadHdr( )** to see if there was an error (indicated by a negative return value); if there was an error, we do not continue with the function. Of course, the most common error would be "File Not Found." We could have added an **RtxPrintf( )** statement to tell about an error before returning.

Because we want to save the menu at the top of the screen when we load images (and some images might be full-screen), we test the height of the image file to make sure it will fit vertically into the work bitmap. If it is too large, we reduce the height value before making the next call. Then we call **GrImLoad( )**, which actually loads the image into the bitmap we specify. The parameters to **GrImLoad( )** include the file name, a pointer that is reserved by DVI for some future use and is passed here as NULL, the target bitmap, and a rectangle in the target bitmap to load into. If the target rectangle is smaller than the image file, only part of the image file will be loaded, taken from the upper left corner of the image. Similarly, if the target rectangle is larger than the image file, the file will load into the upper left corner of the target.

It is possible to specify a target rectangle which runs outside of the target bitmap. In such a case the image would still be loaded, but the runtime software will try to truncate it, however the resulting image may not necessarily be what you expect. Proper practice is to build checks for such errors into your code so that you can deal with them in your own way, and your application will not be messed up and slowed down because the runtime software is trying to deal with errors.

DVI Image Manipulation Functions

We will go into some depth about image and video manipulation in Chapter 13; however, here we will discuss some of the image manipulation (or image processing) primitive functions. We have already discussed **GrCopy( )**, a generic function which can be used for many kinds of manipulations or animations. However, there are other functions in the runtime software to perform different operations, and there will be many more as the microcode library grows. Here are some of the present ones:

- **GrBlend** — performs a dissolve or mix between two source images while copying the result to a destination bitmap

- **GrBrightness** — adds (or subtracts) a constant to the luminance value of every pixel while copying a specified rectangle

- **GrContrast** — increases or reduces the contrast of the luminance value of every pixel while copying a specified rectangle

- **GrMono** — converts each pixel to monochrome (black and white) while copying a specified rectangle

- **GrMosaic** — enlarges pixels by a specified amount while copying a specified rectangle

- **GrSaturation** — adjusts the color saturation value of each pixel while copying a specified rectangle

- **GrScale** — copies one rectangle into another rectangle of a different size, in the same bitmap or between different bitmaps. It does this by *scaling* the first rectangle to fit the second rectangle, interpolating pixel values to produce a smooth result. The scale factor can be different horizontally and vertically — the source and destination rectangles may have different aspect ratios. It also includes an optional filter function which often will make the scaled result look better.

- **GrTint** — adjusts the color value of each pixel to change the tint while copying a specified rectangle

- **GrPolyWarp** — (we have already described) a general function for mapping one polygonal shape into any other polygonal shape having the same number of vertices.

Figure 10.5 shows some of the effects obtained by the use of the above image processing functions in a static mode. Dynamic effects are obtained by calling the functions repeatedly in a loop while changing the parameters according to some kind of algorithm.

Scaling an image
Scaling (changing the size) of an image is an important function which has many uses. Often we will digitize an image in a full-screen high-resolution format and later wish that we had it in a different size in order to display it in a window or to use it on a screen of lower resolution

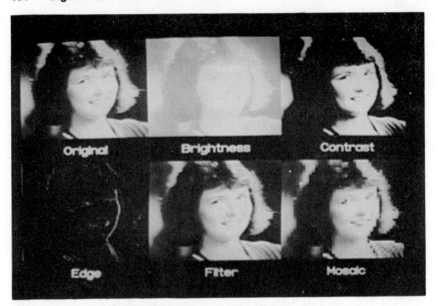

Figure 10.5   Image processing functions

without cropping it. The answer to these problems is a scaling capability. In DVI technology, that need is filled by the **GrScale( )** function; we will now look at a code example which modifies the image we used in the previous example (menu item 5). This is added to the example program **demo.c** as menu item 6. (In fact it will scale any screen which is displayed by **demo.c**.) The source code for the **do_scale( )** function is shown by Listing 10.2.

The first step in the function is to declare local variables for width and height. Then, since we want to scale whatever is on the screen, the drawing region of the screen below the menu is copied to the work bitmap with a call to **GrCopy( )**. Scaling will take place from the work bitmap back to the screen.

We will perform scaling with the filter option turned on, which is done by setting GrBmCopyFilter to TRUE with **GrBmSet( )**. We will scale three times, reducing the size of the image by 100 pixels horizontally and 75 pixels vertically each time. We will locate the scaled output at the upper left corner of the screen below the menu for each operation, so we will get a series of partially overlapping rectangles. A **for( )** loop is set up to do the operation three times. Each pass through the loop then calls **GrScale( )**. The parameters are the bitmap handle for the destination rectangle, followed by the size of the destination rectangle.

Therefore this rectangle is set up to reduce in size for each pass through the loop. The remaining parameters define the source bitmap and the source rectangle, which remains the same for every pass through the loop.

```
/************** Start of Listing 10.2 *************/
/* Scale the current screen image to 3 different sizes */
void    do_scale()
{
I16     x,width = SC_WIDTH,height = 384;

/* First copy the current screen to BMw         */
GrCopy(pBMw, 0, 0, width, height, pBM, 0, MENU_Y + MENU_HEIGHT);

/* Turn on the scaling filter */
GrBmSet(pBM, GrBmCopyFilter, TRUE);

/* Now scale three times between pBMw and pBM,
 * reducing the size each time     */
for (x = 0; x < 3; x++) {
        GrScale(pBM, 0, MENU_Y + MENU_HEIGHT, width - 100,
                height - 75, pBMw, 0, 0, SC_WIDTH, 384);
        width -= 100;
        height -= 75;
        }
GrBmSet(pBM, GrBmCopyFilter, FALSE);
}
/************** End of Listing 10.2 **************/
```

The result of this program is shown by Figure 10.4(b) (see color plates). Note that the **GrScale( )** function works just as well for enlarging an image; however, in that case the resulting image will become pixellated if no filtering is used, or it will just be fuzzy when filtering is turned on. The **GrScale( )** function is slowed down by using the filter function; for the three-pass example described here, it takes approximately 2.5 seconds to complete all three scalings, whereas without filtering the execution time is only 0.6 second. The effect of the filtering is difficult to see when reducing a very busy image like the tulips scene, but aliasing would become visible if an image with sharp diagonal or curved edges was processed without filtering. In the case of enlargement, the filter is usually desirable to reduce visibility of pixellation.

Points made by the software example

In the preceding software discussion, we covered the DVI image file format in detail; this showed that DVI image files are very flexible today, yet they are also expandable to include new features in the future. The software example itself demonstrated the techniques for loading an image and scaling it to any size. Lastly, we listed all of the image processing functions currently available in the DVI microcode library. The use of these functions will be discussed further in Chapter 14.

Need for Still Image Compression

The need for still-image compression is evident in the example just described because we had a single image file which was 412,000 bytes, and it took almost 4 seconds to load it from hard disk. Clearly, reducing these numbers by even a small factor like 2 or 3 can be significant. Such degrees of compression are possible with very little image degradation (sometimes none), and greater degrees of compression are practical for many images which can stand a little compromising. With the 82750 PA, decompression of still images can be done in a fraction of a second, so almost all of the load time reduction obtained by having a smaller file size can be realized in an application. In the next chapter we will cover more about still-image compression.

Summary

Still-image systems have many uses, and a lot of hardware and software is available for still imagery on PC platforms. However, much of the available equipment has limited capability for image manipulation and is slow because there is no high-speed video processor. DVI technology overcomes this with the 82750 PA, which can rapidly perform complex image manipulation algorithms such as scaling and warping. As we will see in the next chapter, the PA also is an outstanding tool for image compression and decompression, and it can greatly enhance a still-image system for both image storage capacity and speed of access.

11

Video Compression
Technology

The desirability of video compression should be clear from the discussion in the previous chapters. Reducing the amount of data needed to reproduce images saves storage space, increases access speed, and it is the only way to achieve digital motion video on personal computers. Therefore, video compression is a crucial element in reaching a fully capable digital video system. This chapter will survey techniques for video compression, and then it will present the still-image compression techniques now available in DVI Technology.

Any video compression scheme can be assessed in four ways:

How much compression is achieved?

How good is the picture?

How fast does it compress or decompress?

What (hardware and software) does it take?

Taking the first question — "how much compression?" — one good way to speak of the amount of compression is to talk about the *number of bits per displayed pixel* needed in the compressed bit stream. For example, if we are displaying a 256x240 (pixel) image from a 15,000-byte bit stream, we are compressing to

$$\text{(bits)} \quad / \quad \text{(pixels)}$$
$$(15{,}000\text{x}8) \, / \, (256\text{x}240) = 2 \text{ bits per pixel}$$

We will use this measure throughout the following discussion to talk about the degree of compression. However, another measure of compression often used is the compression ratio, or the ratio of input data to output data for the compression process. This measure is a dangerous one unless you are careful to specify the input data format in a way that is truly comparable to the output data format. For example, in the example above, the compression process might have used a 512x480, 24 bpp image as the input to the compression process. In that case, the input data was 737,280 bytes, and this would give a compression ratio of 737,280/15,000 = 49. However, because the output display is only 256x240, we achieved 4:1 of that compression by reducing the resolution, so the compression ratio with equal input and output resolutions is more like 12:1. A similar argument can be made for the bits per pixel relationship between input and output — the output quality may not be anything near 24-bpp! The bits-per-displayed-pixel approach avoids all of that confusion; and it will be necessary to evaluate the output picture quality anyway.

Moving to picture quality, that is a much more difficult issue. However, we can divide the world of compression into two parts — *lossless* compression and *lossy* compression. Lossless compression means that the reproduced image is not changed in any way by the compression/ decompression process; therefore, we do not have to worry about the picture quality for a lossless system — the output picture will be exactly the same as the input picture. You may wonder how lossless compression is possible — it is simply the use of more efficient methods of data transmission than the pixel-by-pixel PCM format that we start with from a digitizer.

On the other hand, lossy compression systems by definition do make some change to the image — something is different. The trick is making that difference be hard for the viewer to see. Lossy compression systems may introduce any of the digital video artifacts discussed in Chapter 3, or they may even create some unique artifacts of their own. None of these effects is easy to quantify, and final decisions about compression systems or about any specific compressed image will usually have to be made after a subjective evaluation — there's not a good alternative to looking

at test pictures. The various measures of analog picture quality —
signal-to-noise ratio, resolution, color errors, etc. may be useful in some
cases, but only after viewing real pictures to make sure that the right
artifacts are being measured. In the following discussions, we will not
attempt to put any numbers on our compressed images, but we will show
you what they look like with photographs.

Compression and decompression speed are two separate measures
of compression system performance. In most applications, these two
tasks will be done at different times; they may even be done with totally
different systems at different locations. The reason that compression
and decompression do not happen together is that there is usually
storage or transmission of the image in between the two processes —
storage or transmission is the reason for needing compression in the first
place. Also, in most cases of storing still images, compression speed is
less critical than decompression speed — since we are compressing the
image ahead of time to store it, we can usually take our time in that
process. On the other hand, decompression usually takes place while the
user is waiting for the result, and speed is much more important. (Note
that for motion video compression there is a need for fast compression in
order to capture motion video in real time as it comes from a camera or
VCR—this will be covered in Chapter 12.) In any case, compression and
decompression speed is usually easy to measure and specify.

The question of hardware/software for compression and decompression is mostly a function of the complexity of the algorithms. Very simple
algorithms can run on simple hardware and will not take much software.
Complex algorithms, performing tens, hundreds, or even thousands of
operations relative to every input pixel, will require proportionally more
powerful hardware and software, or they will become extremely slow.
The trade between speed and hardware processing power is an important choice in designing a compression/decompression system.

Another trade is available between the use of general-purpose or
special-purpose hardware. Usually a single algorithm could be implemented more economically in special-purpose hardware which had the
algorithm built right in (called *hard-wiring*) to the circuits. On the other
hand, this kind of system could be much more expensive to design
compared to utilizing some general-purpose hardware already available
off the shelf. Also, the hard-wired system is limited to the one algorithm
and any options which were designed initially — improvements will not
be possible without changing the hardware design and possibly obsolet-
ing all existing equipment. A better approach is to use software-

controlled (programmable) hardware which can potentially implement any algorithm in software. DVI Technology's 82750 PA is one example of low-cost hardware that offers tremendous general-purpose video-processing capability.

Redundancy and Visibility

A digital video image can contain a lot of redundancy — where the same information is transmitted more than once. For example:

- In any area of the picture where the same color spans more than one pixel location, there is redundancy *between pixels*, since adjacent pixels will have the same value. This applies both horizontally and vertically.
- When the scene contains predominantly vertically oriented objects, there is a possibility that two adjacent lines will be the same, giving us redundancy *between lines*.
- When a scene is stationary or only slightly moving, there is a further possibility of redundancy *between frames* — adjacent frames in time are similar.

Compression schemes may exploit any or all of these aspects of redundancy.

Another aspect of displaying a digital video image is that we do not need to display more information than our viewer will be able to see. The best example of this arises from the human eye's poor spatial acuity for certain colors. Because of this characteristic of the viewer, it is not necessary to provide independent color values for every pixel — the color information can be transmitted at lower resolution. This principle is used successfully in NTSC and PAL color television, and it can be utilized in a digital video system as well.

Video Compression Technology

A great deal of research has been done in video compression technology, going back more than 25 years. References 5 and 12 in Appendix D will lead you into the voluminous literature of this research. Many powerful techniques have been developed, simulated, and fully characterized in

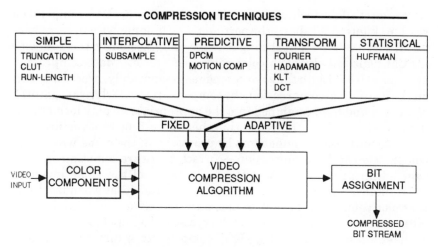

Figure 11.1   Video compression approaches

the literature; in fact, today it is quite difficult to invent something new in this field, it has been so well researched.

However, it should be pointed out that so far there has not been much broad practical application of the more sophisticated video compression approaches outside of the laboratory because of the extreme expense of the hardware needed to implement them in actual products. Most video compression products to date have been directed to the teleconferencing market (group meetings via telecommunications), where equipment costs are in the tens of thousands for each end of a link. That situation is beginning to change through the power of custom VLSI — the DVI Technology system is one example of a low-cost product which can do sophisticated high-quality video compression/decompression.

There are so many video compression techniques that we need a road map in order to keep from getting lost. Figure 11.1 shows a list of some compression techniques. It is not really a map of anything — rather it is a block diagram of how a compression algorithm is applied and which choices are available in constructing an algorithm. With our new powerful video-processing engines, it is possible to make algorithms that use a combination of several different techniques on the same image.

In this discussion, we will use *technique* to refer to a single method of compression — usable by itself, but possibly also used in combination with other techniques. On the other hand, an *algorithm* refers to the collection of all the techniques used by any particular video compression system. Figure 11.1 shows the portfolio of techniques we will cover here.

We will assume that the input to the compression system is always a PCM digitized signal in color component form. Most compression systems will deal with the color components separately, processing each one by itself. In decompression, the components similarly are separately recovered and then combined into the appropriate display format after decompression. Note, however, that there is nothing that requires that the individual color components be processed in the same way during compression and decompression — in fact, there are sometimes significant advantages to handling the components of the same image by different techniques. This brings up immediately that we must choose the color component format to use, and that choice could make a big difference in what is achieved by the system. Two obvious choices that we have already discussed are RGB components or luminance/chrominance components. As each technique is discussed, the significance of the color component choice will also be covered.

Similarly, there is always a possibility of making any technique *adaptive*, which means that the technique can change as a function of the image content. Adaptivity is not a compression technique itself, rather it is a way to cause any given technique to be more optimized locally in the image or temporally in the frame sequence. Almost all the compression techniques we will be discussing can be made adaptive, but of course this adds complexity. Where adaptivity is an important aspect, it will be discussed with each technique.

The output of a compression process is referred to as a *bit stream* — it is usually no longer a bitmap and individual pixels may not be recognizable. The structure of the bit stream is important, however, because it can also affect the compression efficiency and the behavior of the system when errors occur in transmission or storage. Therefore, Figure 11.1 shows a separate box called *bit assignment* — this is where the bit stream structure is imposed on the compressed data. It may be a task which is subsumed in the algorithm, or it may be a separate step in the process.

Simple Compression Techniques

A good example of simple compression is *truncation* — reducing data through arbitrarily lowering of the bits per pixel. This is done by throwing away some of the least significant bits for every pixel. As we know from the discussion in Chapter 3, if we go far enough with truncation, we will begin to see contouring, and our image will start looking like a cartoon. However, many images can stand this up to a point; so, for example, we can usually truncate to 16 bpp with good results on real images. 16 bpp is usually done by assigning bits to color components such as R:G:B 5:5:5, or Y:V:U 6:5:5. In the R:G:B 5:5:5 case, the 16th bit could be used as a flag for some other purpose, such as a keying signal. Truncation is attractive because the compression pre-processing for it is extremely simple.

Another simple compression scheme which creates a different kind of artifact is the color lookup table (CLUT) approach. This is usually done at no more than 8 bpp, which means that the entire picture must be reproduced with 256 or fewer colors at a time. For some kinds of images, that is not as bad as it sounds if the 256 colors are carefully chosen. However, that means each image must be processed ahead of time to choose the 256 best colors for that image (a unique CLUT must be created for each image), and that is a non-trivial amount of pre-processing. Going higher than 8 bpp with CLUT (more colors) will of course give better results, but by the time we get to 16 bpp, it will probably be better to simply use the truncation approach of the previous paragraph because the pre-processing for truncation is so much simpler. (For more discussion of CLUT image quality see Chapter 3 on Color Mapping.)

A third simple technique is *run-length* (RL) coding. In this technique, blocks of repeated pixels are replaced with a single value and a count of how many times to repeat that value. It works well on images which have areas of solid colors — for example, computer-generated images, cartoons, or CLUT images. Depending entirely on the kind of image, RL coding can achieve large amounts of compression — well below 1 bpp. However, its effectiveness is limited to images (or other data streams) which contain large numbers of repeated values — which is seldom the case for real images from a video camera.

Interpolative Techniques

Interpolative compression consists of transmitting a subset of the pixels and using some kind of interpolation to reconstruct the intervening pixels. Within our definition of compression, this is not a valid technique for use on entire pixels because we are effectively reducing the number of independent pixels contained in the output, and that is not compression. The interpolation in that case is simply a means for reducing the visibility of pixellation, but the output pixel count is still equal to the subset. However, there is one case where interpolation is a valid technique — it can be used just on the chrominance part of the image while the luminance part is not interpolated. This is called *color subsampling*, and it is most valuable with luminance-chrominance component images (YUV, YIQ, etc.).

The color components I and Q of the YIQ format were deliberately chosen by the developers of color television so that they could be transmitted at reduced resolution. This is the basis of NTSC color television — a viewer has poor acuity for color changes in an image — so the lower resolution of the color components is really not seen. The same is true for the YUV components, which are used by DVI Technology. YUV was chosen for DVI Technology because YUV fits more into the international standardization of digital television that is now underway.

For example, in a digital system starting with 8 bits each of YUV (24 bpp total), we can subsample the U and V components by a factor of four both horizontally and vertically (a total ratio of 16:1). The selected U and V pixels remain at 8 bpp each, so we still are capable of the full range of colors. When the output image is properly reconstructed by interpolation, this technique gives excellent reproduction of real pictures. The degree of compression works out to 9 bpp:

$$\text{bpp} = \text{(luminance) } 8 + \text{(UV) } 16/\text{(subsamp. ratio) } 16 = 9$$

In the DVI software, 16:1 chrominance subsampling is used in most of the compression algorithms. The 82750 DA contains hardware for doing chrominance interpolation in real time, so images in the subsampled format in VRAM are displayable directly from that format. (An example of the use of 9 bpp mode will be given in Chapter 13.)

Please note that we have used the term "real" images when talking about the advantages of color subsampling and interpolation. It is not as effective on "non-real," i.e., computer-generated images. Sometimes a computer-generated image using color subsampling and interpolation will have objectionable color fringes on objects, or thin colored lines may disappear. This is inherent in the technique; however, the DVI hardware and software have several features which allow mixing of color subsampling and CLUT modes in the same image. This choice can be made on an individual pixel basis, so mixing or overlaying of computer color graphics with subsampled real images can be done without fringes showing on the graphics. Software for doing this is demonstrated in Chapter 13.

Predictive Techniques

Anyone who can predict the future has a tremendous advantage — this applies to video compression as much as it applies to the stock market. In video compression, the future is the next pixel, or the next line, or the next frame. We said earlier that typical scenes contain a degree of redundancy at all these levels — the future is not completely different from the past. Therefore, predictive compression techniques are based on the fact that we can store the previous item (frame, line, or pixel) and use it to predict the next item. If we can identify what is the same from one item to the next, we need only transmit the part that is different because we have predicted the part that is the same.

DPCM

The simplest form of predictive compression is to operate at the pixel level with a technique called *differential PCM* (DPCM). (This is a widely used technique for audio compression — see Chapter 4.) In DPCM, we compare adjacent pixels and then transmit only the difference between them. Because adjacent pixels (usually) are similar, the difference values have a high probability of being small and they can safely be transmitted with fewer bits than it would take to send a whole new pixel. For example, if we are compressing 8-bit component pixels, and we use 4 bits for the difference value, we can maintain the full 8-bit dynamic range as long as there is never a change of more than 16 steps between adjacent pixels. In this case, the DPCM *step size* is equal to one quantization step of the incoming signal.

In decompression, the difference information is used to modify the previous pixel to get the new pixel. Normally the difference bits would represent only a portion of the amplitude range of an entire pixel, meaning that if adjacent pixels did call for a full-amplitude change from black to white, the DPCM system would overload. In that case, it would take a number of pixel times (16, for the example of the last paragraph) before the output could reach full white, because each difference pixel only represents a fraction of the amplitude range. This effect is called *slope overload*, and it causes smearing of high-contrast edges in the image.

ADPCM

The distortion from slope overload may be reduced by going to *adaptive DPCM* (ADPCM). There are many ways to implement ADPCM, but one common approach is to adapt by changing the step size represented by the difference bits. In the previous example, if we knew that the black-to-white step was coming, we could increase the step size before the b–w step came, so that when we got there, the difference bits would represent full range, and a full-amplitude step could then be produced. After the step had been completed, the adaptive circuit would crank the step size back down in order to better reproduce fine gradations. This changes the artifact from slope overload's smearing to *edge quantization* — an effect of quantization noise surrounding high contrast edges. You might have a hard time deciding which is worse!

In the previous example of ADPCM, we glossed over the problem of how the decompression system knows what step size to use at any time. This information must somehow be coded into the compressed bit stream. There are lots of ways for doing that, which we will not go into here, but you should take note that using adaptation with any algorithm will add the problem of telling the decompression system how to adapt. A certain amount of overhead data and extra processing will always be required to implement adaptation.

The DPCM example also highlights a problem of predictive compression techniques in general. What happens if an error creeps into the compressed data? Since each pixel depends on the previous pixel, one incorrect pixel value will tend to become many incorrect pixel values after decompression. This can be a serious problem — a single incorrect pixel would normally not be much of a problem in a straight PCM image, especially a motion image; it would just be a fleeting dot that a viewer might never see. However, if the differential system expands a single dot error into a line that goes all the way across the picture (or maybe even into subsequent lines), everyone will see that! Therefore, predictive compression schemes typically add something else to ensure that recovery from an error is possible and that it happens quickly enough that error visibility will not be objectionable. A common approach is to make a differential system periodically "start over," such as at the beginning of each scanning line or at the beginning of a frame.

After all the previous discussion, it shouldn't be a surprise to say that

2 x 2 ARRAY OF PIXELS

A	B
C	D

TRANSFORM	INVERSE TRANSFORM
$X0 = A$	$A_n = X0$
$X1 = B - A$	$B_n = X1 + X0$
$X2 = C - A$	$C_n = X2 + X0$
$X3 = D - A$	$D_n = X3 + X0$

Figure 11.2   **A simple transform**

DPCM or ADPCM are not widely used by themselves for video compression. The artifacts of slope overload and edge quantization become fatal as we try to achieve more than about 2:1 compression. The techniques, however, do find their way into more complex compression algorithms that combine other more powerful techniques with some form of differential encoding. (DPCM and ADPCM are widely used for audio compression.)

Continuing with predictive compression schemes and moving to the next higher level, we should talk about prediction based on scanning line redundancy. However, line-level prediction is not often used by itself, rather it tends to be subsumed in the two-dimensional transform techniques which very neatly combine pixel and line processing in one package. Since line prediction adds no new concepts to our understanding of compression principles, we will move right into transform coding.

Transform Coding Techniques

A *transform* is a process that converts a bundle of data into an alternate form which is more convenient for some particular purpose. Transforms are ordinarily designed to be *reversible* — that is, there exists an *inverse transform* which can restore the original data. In video compression, a "bundle of data" is a group of pixels — usually a two-dimensional array of pixels from an image; transformation is done to create an alternate form which can be transmitted or stored using less data. At decompression time, the inverse transform is run on the data to reproduce the original pixel information.

In order to explain how a transform works, we will make up a very simple example. Consider a 2x2 block of monochrome (or single color component) pixels as shown in Figure 11.2. We can construct a simple transform for this block by doing the following:

1. Take pixel A as the base value for the block. The full value of pixel A will be one of our transformed values.
2. Calculate three other transformed values by taking the difference between the three other pixels and pixel A.

Figure 11.2 shows the arithmetic for this transformation, and it also shows the arithmetic for the inverse transform function. Note that we now have four new values, which are simply linear combinations of the four original pixel values. They contain the same information.

Now that we have made this transformation, we can observe that the redundancy has been moved around in the values so that the difference values may be transmitted with fewer bits than the pixels themselves would have required. For example, if the original pixels were 8 bits each, the 2x2 block then originally used 32 bits. With the transform, we might assign 4 bits each for the difference values and keep 8 bits for the base pixel — this would reduce the data to only 8 +(3x4) or 20 bits for the 2x2 block (resulting in compression to 5 bits/pixel). The idea here is that the transform has allowed us to extract the differences between adjacent pixels in two dimensions, and errors in coding of these differences will be less visible than the same errors in the pixels themselves.

This example is not really a useful transform — it is too simple. Useful transforms typically operate on larger blocks, and they perform more complex calculations. In general, transform coding becomes more effective with larger block sizes, but the calculations also become more difficult with larger blocks. The trick in developing a good transform is to make it effective with calculations that are easy to implement in hardware or software and will run fast. It is beyond our scope here to describe the transforms that have been developed for image compression, but we can give some of their names so that you can act like you know what they are! One transform which is easy to implement but has limited effectiveness is the *Hadamard* transform. More effective transforms are the *discrete cosine transform* (DCT) and the *Fourier* transform, but of course they take a lot more processing. There are many other more specialized transforms covered in the literature of image compression.

Statistical Coding

Another vehicle for compression is to take advantage of the statistical distribution of the pixel values of an image or of the statistics of the data created from one of the techniques discussed above. These are called *statistical coding* techniques, and they may be contained either in the compression algorithm itself, or they may be applied separately as a bit assignment technique following any algorithm. The usual case for image data is that all possible values are not equally probable — there will be some kind of non-uniform distribution of the values. Another way of saying that is: Some data values will occur more frequently than other data values. We can set up a coding technique which codes the more frequently occurring values with words using fewer bits, and the less frequently occurring values will be coded with longer words. This results in a reduced number of bits in the final bit stream, and it can be a lossless technique. One widely used form of this coding is called *Huffman* coding.

The above type of coding has some overhead, however, in that we must tell the decompression system how to interpret a variable-word-length bit stream. This is normally done by transmitting a table ahead of time, called a *code book*. It is simply a table which tells how to decode the bit stream back to the original values. The code book may be transmitted once for each individual image, or it may even be transmitted for individual blocks of a single image. On the compression side, there is overhead needed to figure out the code book — the data statistics must be calculated for an image or for each block.

DVI Still Image Compression

Because DVI technology is software based, it is theoretically capable of using any of the compression techniques we have discussed. It's just a matter of the right programming! We will describe two still image algorithms available in the DVI software: one is lossless and one is lossy. However, because the DVI compression algorithms are proprietary, we will not discuss the algorithms themselves, only how they are used and how they perform.

Lossless still compression in DVI Technology

The presently available lossless technique applies to 16-bpp images only. (This is just because 16 bpp is the first version which has been developed; there's no reason the same thing couldn't be done for other bpp formats.) You can think about the performance of this algorithm as being like the run-length technique — compression is good for graphics-like images and becomes less effective as the image complexity increases. Using the images from the examples of Chapters 9 and 10, the smiley face (all graphics) will compress about 7:1, whereas the very busy tulips scene (a photograph of a real scene) compresses only 1.2:1. The beauty of this technique is that it is lossless — the output is *exactly* the same as the input.

We are now going to look at a software example for lossless compressed images. (The non-software reader may skip ahead to **Lossy Compression**.) Listing 11.1 is a revision of the **do_image( )** function of Chapter 10 to include lossless decompression for loading the image.

```
/*************** Start of Listing 11.1 ************/
/*  Load either a compressed or uncompressed 16-bit image   */
void    do_image1()
{
I16     x,width,height,comp = 0,ylen,AlgVer;
char    *fileName = "tul";
char    *fName;
char    name[64];
U32     data_addr = 0x78000;
FILE    *f1;
hGrImHdr ImHdr;
hGrFunc Alg;

/*  See if a compressed file exists   */
f1 = StdEnvFOpen(fileName, "rb", ".c16","VIM",fName);
if (f1 > 0) {
```

```
      fclose(f1);
      comp = 1;
      } else {
    f1 = StdEnvFOpen(fileName, "rb",".i16","VIM",fName);
     if (f1 == NULL) {
         RtxPrintf("file %s not found\n",fileName);
          return;
          }
     fclose(f1);
     comp = 0;
      }

/*  Get the size of the image  */
GrImHdrAlloc(&ImHdr);
x = GrImLoadHdr(fName, NULL, pBMw, ImHdr);
if (x < 0) return;
GrImHdrGet(ImHdr, GrImHdrXLen, &width);
GrImHdrGet(ImHdr, GrImHdrYLen, &height);
GrImHdrFree(ImHdr);

/*  If uncompressed, load the image into BMw   */
if (comp == 0) {
   x = GrImLoad(name, NULL, pBMw, 0, 0, width, height);
   if (x < 0) return;
   /*  Now copy the image to the screen, centered */
   GrCopy(pBM, (SC_WIDTH - width)/2, (SC_HEIGHT - height)/2,
          width, height, pBMw, 0, 0);
   } else {    /*  It is a compressed file  */
   /* load code into memory, decompress directly to screen */
   x = GrImLoadCode(name, NULL, data_addr);
   if (x < 0) return;
   /*  Get the algorithm version, pass it on  */
   GrCodeGet(data_addr, GrCodeAlgVer, &AlgVer);
   GrDecodeOpen(AlgVer, 16, &Alg);
   GrDecode(Alg, pBM, GR_PLANE_YVU, (SC_WIDTH - width)/2,
          (SC_HEIGHT - height)/2, data_addr, NULL, NULL, 0);
   }
}
/*************** End of Listing 11.1  ***************/
```

This function is set up to first look for a compressed file (with the name extension .c16), if that is found, the variable **comp** is set to 1; otherwise, the function will go on to look for the same filename with a .i16 extension. The DVI runtime function **StdEnvFOpen( )** is used to look for the files. This function is given a filename without an extension and then is given the Microsoft C flag string for file access — in this case "rb", which means *read* and *binary*. Following that, **StdEnvFOpen( )** can be given a DOS environment variable name for a path where image files may be found. In the DVI runtime software, this environment variable is VIM. The last parameter in **StdEnvFOpen( )** is a character pointer to a buffer where the actual filename with path name will be placed if a file is opened.

The return value of **StdEnvFOpen( )** will be zero if a file was not found and positive if one was found. This is tested to determine whether to set the **comp** variable or to go on and look for an uncompressed version of the file. If a file was opened, it must be closed at this time by a call to **fclose( )**.

We then proceed to load the image file header to obtain the pixel size of the file, the same as in Chapter 10. Then, if the file was uncompressed we load it off-screen and copy it to the screen, again the same as in Chapter 10. However, if the file is compressed, we must decompress it. This is done by loading the compressed bit stream data to an off-screen buffer with **GrImLoadCode( )**, and then decompressing it directly to the screen with **GrDecode( )**.

GrImLoadCode( ) takes the filename, a reserved NULL pointer, and a VRAM address to load the data.

In order to set up for decompression, we first call **GrCodeGet( )** to get the algorithm version number and pass it to **GrDecodeOpen( )**. This sets up the data structures for decoding, loads the proper microcode into VRAM, and passes back a handle to the structures. Following this, we can call **GrDecode( )**, telling it the algorithm handle, what bitmap to decompress into (the screen pBM), the image planes involved, where (x and y) to locate the decompressed image in the destination bitmap, the VRAM address of the compressed data, and several NULLs (which are used for other algorithms). The reason for reading the algorithm number from the compressed data is so that the same decompression code can handle several algorithms. There are already two lossless 16-bit algorithms — one does compression in a fraction of a second, and the second takes several seconds to compress but achieves about 50% more compression as a result.

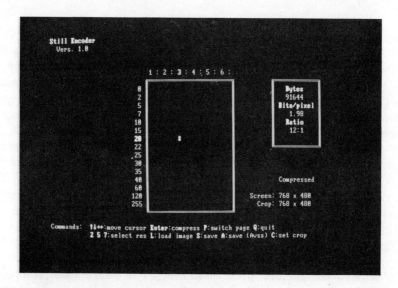

Figure 11.3   Vcomp command screen

The decoding process is not instantaneous — for this example it takes about 1/2 second. This gives a vertical wipe effect as the decompressed image becomes visible on the screen. If this is objectionable, it would be necessary to have more than 1 megabyte of VRAM so that the decompression could be pointed to another off-screen bitmap. Then the image could be copied from that bitmap to the screen to give an almost instantaneous visual transition. With 1 megabyte VRAM there would not be enough space to keep both the compressed code and the decompressed image off-screen.

Lossy compression

If more compression is needed than that provided by the lossless algorithm, or if one wishes to work with 9-bpp images (which themselves are almost 2:1 smaller than 16-bpp images), then the lossy compression algorithm must be used. This algorithm is implemented in the **vcomp** program, which is a development tool for creating lossy-compressed images. The program provides user control of two parameters to make tradeoffs between the degree of compression and the image quality. Compression in the range of 1–2 bpp can deliver good images with this algorithm.

The user interface screen of **vcomp** on the standard output monitor is shown in Figure 11.3, and Figure 11.4 (see color plates) shows some sample compressions of a 768x480 original image.

The original image must be in 9-bpp format, and it can be any size up to 768x480 resolution. The screen size is selected by the number keys — 2 for 256x240, 5 for 512x480, and 7 for 768x480. An image is loaded by typing "L" which brings up a dialog box for a filename. This image is displayed in a minimum compressed format, but the original may be viewed by typing "P".

The box in the center of Figure 11.3 allows user adjustment of the parameters by use of the arrow keys, which move the # symbol around the box. To the right and down results in greater compression. The horizontal axis is something like truncation, and the vertical axis is something like filtering. Hitting the return key will cause compression to be done with the current parameter choice and the result immediately displayed. This can be done as many times as needed to evaluate the range of compression for any particular image.

When a satisfactory compressed image has been chosen, the result may be saved by typing "S", and entering a filename into the dialog box. This will create three compressed image files with filename extensions .cmy, .cmi, .cmq. An alternate save mode (type "A") is also available to save the image in AVSS file format with extension .avs. This image can be displayed by the AVSS software, which is described in Chapter 12.

Standardization of Algorithms

Driven by telecommunication-based needs for video compression, there is a strong effort to develop international standards for video compression algorithms. This has already been under way for several years in the ISO (International Organization for Standardization) and the IEC (International Electrotechnical Commission). While this would not appear to be very important to PC-based video compression which is still evolving and is probably not ready to settle on only one algorithm, it becomes important when you look for the broadest possible application of the compression hardware (chips), and for the systems (PCs) which use that hardware. In the long run, multimedia PCs will need to communicate with each other and this will require standards — internationally. Therefore, it makes sense that video compression for PCs and video compression for telecommunication should get together.

DVI Technology has a strong presence in this activity in order to be sure that the Technology, which is software-based and should theoretically be able to run any algorithm, will actually do that for telecommunications algorithms which are under consideration and will become standards in the future.

Without explaining the complete organizational structure involved, there are two working parties for algorithm standardization in a joint ISO-IEC Committee (called JTC1). These working parties are the JPEG (Joint Photographic Expert Group), which is considering still image standards in several categories, and MPEG (Motion Picture Coding Expert Group), which is considering algorithms for motion video compression. These groups are working at a timetable which should lead to some results in late 1990 or early 1991. However, the field of image and video compression is complex and fast-moving and there are over 40 companies participating in the standards work, so it is likely that standards in this field will evolve over a number of years.

Summary

The general techniques of video compression have been described, and the capabilities of the DVI system were covered in detail. The wide variety of compression techniques already available and the large amount of research and development still going on in this field means that a software-based video system is desirable in order to keep up with the state of the compression art as it continues to grow.

12

Digital Motion Video Systems

In the still-image compression systems that we discussed in Chapter 11, we really gave very little consideration to the matter of compression or decompression *speed*. With still images, our processes only needed to be fast enough that the user did not get bored waiting for things to happen. However, when one begins to think about motion video compression systems, the speed issue becomes overwhelming. Processing of a single image in one second or less is usually satisfactory for stills. However, *motion video* implies a high enough frame rate to produce subjectively smooth motion, which for most people is 15 frames per second or higher. *Full-motion* video as used in this book refers to normal television frame rates — 25 frames per second for European systems, and 30 frames per second for North America and Japan. These numbers mean that our digital video system must deliver a new image every 30–40 milliseconds! If the system cannot do that, motion will be slow or jerky, and the system will quickly be judged unacceptable.

At the same time that we need more speed for motion compression, we also need to accomplish more compression. This comes about because of *data rate* considerations. Storage media have data rate limitations, so they cannot simply be speeded up to deliver data more rapidly. For example, the CD-ROM's continuous data rate is fixed at 153,600 bytes per second — there is no way to get data out faster. If CD-ROM is being used for full-motion video at 30 frames per second, we will have to live with 5,120 bytes per frame. Therefore, we face absolute limits on the amount of data available for each frame of motion video (at least on the average); this will determine the degree of compression we must achieve.

For CD-ROM at 5,120 bytes of data per frame (40,960 bits per frame) and at a resolution of 256x240 pixels, the required compression works out to be 0.67 bits per pixel. Some still compression systems can work down to this level, but the pictures are not very good, and 256x240 already is a fairly low pixel count. Therefore we should look at motion video to see if there are possibilities for compression techniques which can be used *in addition* to the techniques we discussed for stills.

Fortunately, motion video offers at least one opportunity to achieve additional compression beyond what can be done with stills. That is the redundancy between adjacent frames — a motion video compression system can (or must) exploit that redundancy. Such techniques are called *motion compensation*, and they are predictive compression at frame rate. We will discuss motion compensation shortly.

Another concept that comes into play with motion video systems is the idea of *balance* between compression and decompression. A balanced compression/decompression system will use the same hardware for both compression and decompression and perform both processes at roughly the same speed. Such a system for motion video will require hardware that is too expensive for a single-user system, or else it will have to sacrifice picture quality in favor of lower cost hardware. However, this problem can be effectively bypassed by the use of an *unbalanced* system where the compression is performed on expensive hardware, but the decompression is done by low-cost hardware. This works in situations where the single-user system needs only to play back compressed video which has been prepared ahead of time — it will never have to do compression.

In fact, most interactive video applications do not require that the end-user system contains a compression capability — only decompression. Motion video for this class of application can be compressed (once) during the application design process and the final user only plays back the compressed video. Therefore, the cost of the compression process is shared by all the users of the application. This concept can lead further to the establishment of a centralized compression service which performs compression for many application developers, thus sharing the costs even further .

DVI Technology can do both balanced and unbalanced motion video compression/decompression. The unbalanced approach is called *Production-Level Video* (PLV) — video for PLV must be sent to a central compression facility which uses large computers and special interface equipment, but any DVI system is capable of playing back the resulting compressed video. The picture quality of PLV is the highest that can currently be achieved. The other DVI compression approach is called *Real Time Video* (RTV) — it is done on any DVI system that has the ActionMedia Capture Board installed. Playback of RTV is on the same system or any other DVI system. Because RTV is a balanced approach which requires that compression be done with only the computing power available in a DVI system, RTV picture quality is not as good as PLV picture quality.

Motion Compensation

Consider the case of a motion video sequence where nothing is moving in the scene. Each frame of the motion video should be exactly the same as the previous one. In a digital system, it is clear that all we need to do is transmit the first frame of this scene, store that and simply display the same frame until something moves. No additional information needs to be sent during the time the image is stationary. However, if now a dog walks across our scene, we have to do something to introduce this motion. We could simply take the image of the walking dog by itself, and send that along with the coordinates of where to place it on the stationary background scene — sending a new dog picture for each frame. To the extent that the dog is much smaller than the total scene, we are still not using much data to achieve a moving picture.

The example of the walking dog on a stationary background scene is an overly simplified case of motion video, but it already reveals two of the problems involved in motion compensation:

How can we tell if an image is stationary?
How do we extract the part of the image which moves?

We can try to answer these questions by some form of comparison of adjacent frames of the motion video sequence. We can assume that both the previous and the current frames are available to us during the compression process. If we do a pixel-by-pixel compare between the two frames, the compare should produce zero for any pixels which have not changed, and it will be nonzero for pixels which are somehow involved in motion. Then we could select only the pixels with nonzero compares and send them to the decompressing system. Of course we would have to also send some information which tells the decompressing system where to put these pixels.

However, this very simple approach, which is a form of frame-to-frame DPCM, is really not too useful because of several problems. First, the pixel compare between frames will seldom produce a zero — even for a completely stationary image, because of analog noise or quantizing noise in the system. This could be gotten around by introducing a threshold that would let us accept small comparison values as zero, but there is a more serious problem — images from video or film cameras are seldom stationary. Even if the scene itself contains no motion (which is unusual in natural scenes) the camera may be moving slightly, causing all pixel compares to fail. Even partial pixel movements will create changes large enough to upset the comparison technique.

Therefore, more sophisticated techniques are needed to do the motion detection for the purpose of motion compensation. This problem is usually addressed by somehow dividing the image into blocks, just as we did with still images for transform coding. Each block is examined for motion, using approaches which consider all of the pixels in the block for motion detection of that block. If the block is found to contain no motion, a code is sent to the decompressor to leave that block the way it was in the previous frame. If the block does have motion, a transform may be performed and the appropriate bits sent to the decompressor to reproduce that block with the inverse transform.

If enough computing power is available for the compression process, still more sophisticated approaches can be pursued. For example, blocks which contain motion can be further examined to see if they are simply a translation of a block from the previous frame. If so, only the coordinates of the translation need to be sent to tell the decompressor how to create that block from the previous frame. Even more elaborate techniques can be conceived to try to create the new frame using as much as possible of the information from the previous frame instead of having to send new information.

DVI Production-level Compression

PLV compression is an unbalanced approach where a large computer does the compression and the DVI system does the decompression. It takes a facility costing several hundred thousand dollars to perform PLV compression at reasonable speeds. Since this cost is too much for a single application developer to bear, centralized facilities are being provided where developers can send their video to be compressed for a fee. We will discuss what such facilities involve, and how they will be made available to developers.

High-quality motion video compression has difficulties right at the start — the data rates created by the initial digitizing are high, even for large computers. This happens because the initial digitizing really has to be done in real time to obtain the best quality. In most cases, the input video medium for compression will be an analog video tape — for best results, it will be 1-in. broadcast quality tape. Although 1-in. tape machines can play at slow speeds, they do it by introducing frame storage and processing which would interfere with the quality of the compressed result. The only way to get around that processing is to run the VTR at normal play speed — 30 frames per second. Therefore, for best quality, we must invest in digitizing and interface hardware which will let the VTR run at normal speed and capture the digital data on computer disk.

For PLV compression, the real-time video from the VTR is digitized, filtered, and chrominance subsampled before storage on digital disk. This still requires a data rate of about 2 megabytes per second — a 1.2 gigabyte digital disk only holds 10 minutes of this partially compressed digital video. Then, in non–real time, the data is taken frame by frame from the digital disk and run through the PLV compression algorithm.

The compression algorithm is running on a Meiko parallel processor CPU, which has 64 Transputer processing nodes. Compression takes about 3 seconds per frame on this machine — still about 90 times slower than real time. (A minute of final compressed video will take 90 minutes to compress.) This is the speed of the first installation; it will be made faster by improvement of the software and by expanding the Meiko to more computing nodes as the compression business volume grows. Additional facilities will also be set up worldwide as the need develops.

PLV performance
The DVI PLV compression algorithm is proprietary and will not be described here. Its performance is also difficult to describe or show here, because it does not make sense to show still frames from a motion sequence. This is because an individual frame from motion video may

contain artifacts which are not visible when those frames are delivered to a viewer at 30 frames per second. There is a significant degree of visual averaging taking place when viewing 30 frames per second video. This is also true for normal analog television — noise artifacts become highly visible in a single still frame, whereas the averaging between frames in a motion sequence makes noise much less visible. You can observe this problem if you experiment with a VCR which has slow-motion or still-frame features. A stopped picture looks much worse than normal-motion pictures. Anyway, PLV compressed video delivers full-screen, full-motion pictures at a quality subjectively competitive with 1/2" VCR pictures.

The PLV compression algorithm must be given goals for the size of the compressed bit stream and for the amount of 82750 PA time per frame which will be devoted to decompression. Even when working with CD-ROM, we will often want to use fewer than 5,120 bytes per frame for the average data rate of the video because we want to leave space in the CD-ROM rate for audio or possibly other data. We also may wish to display the motion video at less than full-screen in order to save data so that more than 72 minutes could be on one CD-ROM disc. In that case, we would specify a cropped picture to be compressed to fewer bytes per frame.

Another way to effectively reduce the compressed data rate is to lower the video frame rate. In some cases, 15 frames per second would be fast enough — this can be used either to cut the video data rate in half or to allow more than 5,120 bytes per frame to achieve somewhat higher video quality.

In the case of the 82750 PA decompression time, there are 33 milliseconds per frame available at 30 frames per second (40 milliseconds per frame at 25 frames per second), but we may not want to let all of that time be used for decompression because we need the PA to perform some other processes on each frame, such as drawing graphics over the motion image.

You can see that there is a multidimensional tradeoff here involving four interacting parameters, which together will determine the resulting picture quality:

- Image cropping (pixel count)
- Compressed data bytes per frame
- Video frame rate

- Decompression processing time

The PLV compression process takes all of these parameters as input and it will try to produce the best quality pictures within these constraints.

Because PLV is a frame-to-frame compression scheme, there are some special considerations involved in starting up a scene or in starting a scene in the middle. The first frame of a motion sequence must be treated as a still image (called a *reference frame*), and additional time is required to send all the data for a reference frame, which is about three times the data of an average motion video frame. (If we are using motion video at 5,120 bytes average per frame, a reference frame will require around 15,000 bytes of data.) If it is intended that a scene will be started by the application at several different points, it is further necessary to introduce reference frames at those points. This can often be done without causing any noticeable interruption of motion when the scene is played from end to end, because the DVI decompression software uses multiple frame buffers in VRAM so that variations in the input compressed data rate can be accommodated without affecting the displayed frame rate. In any case, if you have special needs for reference frames in your video, they have to be expressed at the time you order PLV compression.

DVI Real Time Compression

The use of a centralized compression service that is remote from the developer of an application introduces delay and expense into the application development process. In creating an application, a developer needs a way to experiment with his or her video and audio in the context of the application without incurring this delay and expense. This need is filled by DVI's Real Time Video (RTV), which is a compression process that is done in real time on a DVI development system. With RTV, the developer may compress his or her video and audio to the same file size as from the PLV service, and then use those files in the application under development in exactly the same way that the final PLV files will be used. By this means the developer may experiment as much as needed and actually try out the complete application *before* sending any video out for PLV compression.

The tradeoff in RTV is picture quality. To accomplish motion video compression with only the resources of a DVI development system means that RTV is lower in resolution and lower in frame rate than PLV. Compression is done with 82750 PA microcode and, while the PA is very powerful among its peers, in the milliseconds available to compress a frame in real time, the PA does not compare with the computer cycles available in 3 seconds on a Meiko parallel machine. Therefore, the algorithm must be simplified. However, the results produced are good enough to fill most needs for application development and testing. For some applications which do require real-time compression within the application, RTV may completely fill the bill.

RTV compression allows the user to make some trades of compression vs picture quality if the RTV-compressed code will never have to be stored on a CD-ROM. By allowing the data rate to go higher than 153,600 bytes per second using fast hard disk storage, the RTV frame rate can be increased to 30 frames per second. The Intel RTV capture tool allows several choices of this tradeoff.

Communication between RTV and PLV occurs through the medium of SMPTE time code. The original 1-in. video tape which will eventually go to PLV compression must have SMPTE time code on it. When this tape is compressed by RTV for development purposes, the time code is captured for storage with the video frames. RTV is not frame-to-frame compression, so an RTV file can be started or stopped at any point. In the RTV mode, decisions about in and out cut points for the displayed video can be made and the time code values may be read from the RTV file data to create the edit list which will be used for the PLV compression. After all decisions about video material have been made, the master 1-in. tapes and the edit list go to the PLV compression facility for final compression of exactly and only the selected scenes.

Software for DVI Motion Video

Playback of DVI motion video and audio — whether RTV or PLV — is handled by the runtime software module called the Audio/Video Support System (AVSS, pronounced "avis"). AVSS requires that the video and audio data be in a special format called the AVSS file format. We will look at that in some detail. We will then follow with a software example of using AVSS to play motion video. AVSS is capable of playing motion video, motion video plus audio, audio alone, or video stills. It also has

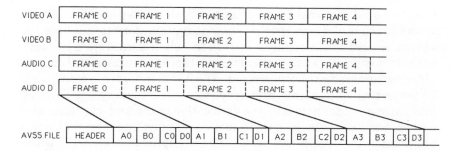

Figure 12.1   AVSS interleaving of multiple streams into one file

features to allow the application programmer to do other operations at the same time as playing video or audio, either in synchronism with what is happening in the video or audio, or independently.

AVSS file format
Just like the image file format we discussed in Chapter 10, the AVSS file format (filename extension .avs) provides for a lot of flexibility and expandability. However, it is much more complex than the single-image files, so we will just talk about AVSS files at the concept level. AVSS files are designed to support multiple simultaneous *streams* of data — for example, one AVSS file might contain one video stream and two audio streams for stereo audio. In a more complex example, one AVSS file might contain four video streams and also four audio streams — one for use with each video stream. This kind of combination would be needed if the application wanted to play any one of four video streams and instantly switch to another one while playing.

The multiple-stream-per-file approach is necessary because a slow-access storage device like CD-ROM can continuously play only one file at a time. (The seek delays of trying to play two different files in parallel from a CD-ROM would either require massive buffering or the average data rate would seriously slow down — either is intolerable.) Therefore, we must have the ability to put things that need to play simultaneously into the same file.

In addition to the audio and video stream types, three additional stream types are provided: underlay data, data, and image. *Underlay data* is any kind of data which must be provided in real time along with the audio and video. (An example of underlay data is the time code which must be stored with each frame of an AVSS file created by RTV.) A *data stream* consists of a sequence of bits that represent computer data but will not be used by the application until the whole data stream has been stored in memory. An *image stream* consists of a sequence of still images, either compressed or uncompressed. The AVSS software is responsible for converting video, audio, and image streams into images and sound. AVSS also provides access by the application to the underlay data as it arrives, and it collects the data from data streams into memory buffers identified by the application, notifying the application when the whole data stream has arrived.

The storage medium for holding an AVSS file (CD-ROM, hard disk, RAM disk) of course delivers only a single stream of bits while a single file is being read. The multiple streams that go into an AVSS file must be *interleaved* into a single bitstream — this is done on an *AVSS file frame* basis, and it is shown by the diagram of Figure 12.1. An AVSS frame may or may not be the same as a video frame — it is a bundle of data which contains one block from each active stream. Normally one video block in an AVSS frame would be a complete definition of a single image or a video frame (although a video frame may be frame-to-frame compressed, which assumes that a previous frame is already available), but the file format could (if necessary) also handle a situation where one video block did not represent a complete image.

The individual blocks which go into an AVSS frame may not be of a constant size. Particularly with compressed video, some frames may need more or less data than others. However, it is important with slow-access devices like CD-ROM to maintain a certain average playback rate so that a long sequence can be played without the CD-ROM having to do a seek. A seek would introduce at least a one-revolution latency of several hundred milliseconds, and maybe more, and would probably cause an unacceptable interruption of data. Therefore, AVSS files often contain *padding*, which is simply an additional stream of dummy bits interleaved with the others to control the average byte rate.

Operation of AVSS

AVSS operates through the DVI real-time executive module, RTX. We need to introduce some of the RTX concepts in order to explain AVSS. RTX is a form of multitasking which runs inside of a DVI application. The basic working unit of RTX is a *task*; the interrupt structure of the PC is set up so that a number of RTX tasks can operate "simultaneously." This is accomplished by interrupting the CPU thirty times per second

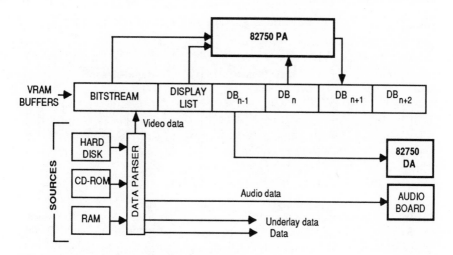

Figure 12.2   AVSS data flow diagram

(not synchronized with the video frame rate, however). Depending on the priority structure for the tasks which have been started by the application, the RTX scheduler will give each task a chance to run in sequence. Tasks can communicate with each other through a system of *event flags*.

When AVSS is running it activates three high-priority RTX tasks: a *server* task which handles data transfer from mass storage into memory, a *Decode* task which handles control of the 82750 PA, and a *Display* task which handles display of image frames by the 82750 DA. These three tasks do all the detail work of playing video and audio.

Figure 12.2 shows a diagram of the data flow associated with AVSS. Data is read in from storage at the bottom left, either hard disk, CD-ROM, or RAM. The first step is to parse the data into the individual streams: video data goes into VRAM, audio data goes to VRAM on the audio board, and underlay and data go to buffers that the application sets up in the system RAM. The server task does all this.

The center of Figure 12.2 shows what AVSS is doing with VRAM. At the left is the input buffer where the compressed video bitstream data is stored. Next to the right is an area of VRAM used for the 82750 PA display list. (This is the same space used by the graphics software, and it usually is located at the top of VRAM.) Then to the right is a number of display buffers for decompressed video frames (six is a typical number,

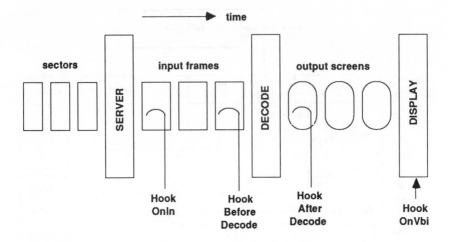

Figure 12.3   AVSS process flow and hook routine positioning

although some situations may require even more). The Decode task of AVSS takes care of telling the PA to run the decompression on the bitstream data and place the result in a particular buffer — in the figure, the PA is decompressing to buffer DB_{n+1}. Of course, the PA also has access to the previous frame DB_n and it is getting its commands through the display list. At the same time as this, the Display task has told the DA to display the DB_{n-1} frame, which had previously been decompressed. Therefore, there are several buffers in use ahead of the one currently being displayed.

To make it possible for the application program to command other things to happen while AVSS is running, there is the concept of an AVSS *hook routine*. Hook routines are C functions which the application can set up to be called at particular points in the AVSS processing for each frame of video. Figure 12.3 shows four points where hooks can be placed in the pipelined processing of AVSS. The diagram shows how frames of video data move through the system, from left to right. The actions of the three AVSS tasks are represented by the blocks Server, Decode, and Display. At any moment in time, there will be something stored in the buffers ahead of each of these tasks. Hook routines can be set up to be called at the points in the cycle indicated by the hook symbols. The first hook (HookOnIn) is called when a complete frame of data has been loaded into an input buffer. The second hook point (HookBeforeDecode) is called when any frame of compressed data is about to begin decompression. If a hook is activated there, the identified routine will be called and passed the frame number and the buffer location of that frame.

The third hook point (HookAfterDecode) is immediately after decompression of a frame has been completed. This is normally where a routine to draw graphics in the buffer would be inserted. The last hook point (HookOnVbi) is called for each vertical blanking interval (VBI). This would be used to do something which had to be closely timed with the display of the frame, such as drawing a cursor in accordance with mouse or joystick position. If this were done at HookAfterDecode time, there would be a delay until that frame was displayed. This could be very bad if the playback frame rate is low.

The CPU cycles or 82750 PA cycles that will be used by hook routines must be taken into account in planning an application. The PA will be fully loaded by decompression unless the compression service was told to compress to use a shorter decompression time. If extensive PA processing by hook routines is expected, then an appropriately shorter decompression time goal must be given to the compression service. Since shorter decompression time means a tradeoff in picture quality, such a choice should be approached carefully to be sure that the pictures will be acceptable. Another consideration is that PA display list commands generated by hook routines should not get out of step with the display list commands generated by the Decode task of AVSS. The placing of hook points has been planned to take that into account, and it should not be a problem unless hook routines have an excessive processing time before issuing PA commands.

During AVSS playback, the CPU is not usually fully loaded, and there are cycles available for other tasks. But again, any plan to use a lot of CPU time while video and audio plays should be carefully evaluated and tested to ensure satisfactory operation.

AVSS Software Example

Listing 12.1 is a complete program for playing an AVSS audio, video, or audio/video file. It is called **av.c**, and is about the simplest implementation of AVSS that is possible. However, it shows many of the AVSS software concepts.

```
/*************** Start of Listing 12.1 ****************/
/*  Simple example program to play an AVSS file         */

#include "vdvi.h"

#define SIZEX 256          /* resolution of screen */
#define SIZEY 240
#define NUMSCR 7           /* Number of screen buffers  */

#define YADDR    0x10000   /* Y address of 1st screen */
#define YSIZE    0x10000   /* Size of each Y plane, rounded up */
#define VADDR    0x80000   /* V address */
#define VUSIZE   0x4000L   /* Size of VU planes, rounded up */
#define DECADDR  0xa0000   /* Decode addr: 64K work area */
#define VIDADDR  0xb0000   /* Video input buffer */
#define VIDSIZE  0x40000
#define AUDSIZE  0x8000    /* Audio input buffer */

main(argc,argv)
int argc;
char **argv;
{
hGrBm bm[NUMSCR];
hAvFile file;
int i;
VAddr y,v;

DviBegin();
GrBmDisplay(NULL);

/*  Define the bitmaps for the screen buffers   *
 *  and clear each one to black                 */
y = YADDR;  v = VADDR;
for (i=0; i<NUMSCR; i++) {
    GrBmAlloc(&bm[i]);
    GrBmSetPlanar(bm[i],GR_BM_9,y,SIZEX,SIZEY+1,SIZEX,v);
    GrBmSetRgb(bm[i],GrBmDrawColor,0,0,0);
    GrRect(bm[i],0,0,SIZEX,SIZEY+1);
    y += YSIZE;  v += VUSIZE;
    }
AvSet(NULL,AvScrArray,bm);  /* tell AVSS about screen buffer */
AvSet(NULL,AvScrCnt,NUMSCR);
AvSet(NULL,AvDecodeAddr,DECADDR);  /* set global decode addr */

if (AvOpen(argv[1],&file) < 0)      /* open file */
        { RtxPrintf("Can't open file\n"); goto ending; }
```

```
AvSet(file,AvFileVidAddr,VIDADDR);  /* set up input buffers */
AvSet(file,AvFileVidSize,VIDSIZE);
AvSet(file,AvFileAudAddr,StdMemAlloc(AUDSIZE));
AvSet(file,AvFileAudSize,(U32)AUDSIZE);

AvCtrl(file,AvPlay);            /* play it */
AvCtrl(file,AvFileWaitDone);    /* wait until done */
AvClose(file);                  /* close it */
GrBmDisplay(NULL);              /* blank screen again */

ending:
DviEnd();
}

/***************** End of Listing 12.1 ***************/
```

The program begins with the normal #includes for a DVI program, and then we define some constants for the VRAM buffer setup. SIZEX and SIZEY define the screen buffer pixel dimensions, and NUMSCR defines that there will be seven buffers. The motion video screen buffers are 9-bit format, so we have to locate both the Y-plane and the VU-planes. YADDR is the VRAM address of the first buffer Y-plane, and the seven Y-planes will be placed together 64k (0x10000) apart. The Y-planes themselves actually take only 256 x 240 = 61,440 bytes, but it is convenient to round this up to 65,536 bytes (64k). Similarly, the VU-planes will be placed above the Y-planes, starting at address 0x8000, and spaced 16k (0x4000) apart. Thus, seven of those will take us up to a little less than 0xa0000.

In addition to the motion video screen buffers, AVSS requires the allocation of a raw data input buffer and a work space in VRAM. We will place the work space (DECADDR) at 0xa0000, allowing 64k, and the input buffer (VIDADDR) at 0xb0000. The input buffer size will be four pages of VRAM or 0x40000. The last thing we will define is the audio buffer size — this will be set up in system RAM, and we will call for 32k (0x8000).

Then we begin the program itself — **main( )**. In this case we will set up for a command line argument by declaring the variables **argc** and **argv**. This will be used to pass the filename of the AVSS file to be played.

In **main( )** we declare a 7-element bitmap array to hold the handles for the seven screen bitmaps, and an AVSS file handle of type hAvFile.

We also will need an integer **i** for counting, and **y** and **v** VRAM address variables.

Now we can call **DviBegin( )** to set up the DVI subsystems, and then a **GrBmDisplay(NULL)** to blank the screen while we set things up. Then we need to allocate and initialize the seven screen bitmap buffers, which is done by placing the allocation and initialization calls into a seven-pass **for** loop. At the same time, we can add a **GrRect( )** call in the loop to clear each buffer to black.

Having set up the screen buffers, we need to tell AVSS about them, which is done with three calls to **AvSet( )**. Then we call **AvOpen( )** to open the AVSS file. If this call fails, we abort the program. If it succeeds, we go on to make more **AvSet( )** calls to tell AVSS about the input and audio buffers we want to use. Note that the audio buffer is allocated from system RAM by a call to **StdMemAlloc( )**.

To actually play the file, we call **AvCtrl( )** with the AvPlay flag. This function returns immediately while playing proceeds. This would be useful if we wanted to do some other things while playing, such as looking for some kind of user input. However, in this case, we don't want that, but rather want the program to wait for the playing to finish and then blank the screen and quit. The way to do that is to call **AvCtrl( )** again with the flag AvFileWaitDone. This call will not return until playing is finished.

The final action in the program is to close the AVSS file and go through a normal ending procedure.

Additional AVSS features and capabilities

AVSS has default values for parameters such as playback frame rate, video positioning on the screen, audio rates, audio volume, audio/video relative timing, etc. However, any of these parameters may be changed before playback or at any time during playback by use of the **AvSet( )** function. By means of a hook routine, it also is possible to change any of these parameters under the control of information contained in an underlay data stream.

If there are enough CPU and 82750 PA cycles, AVSS may be invoked more than once at the same time — two (or more) AVSS files could be played simultaneously. However, this would not be possible from CD-ROM because of the seek time problem we mentioned earlier. It is possible when the two files come from different storage devices. For example, in a surrogate travel application, background audio could be played by AVSS from a RAM disk while another invocation of AVSS plays the video from CD-ROM. The audio used here may have come earlier from CD-ROM as a separate data file or as a data stream going along with the travel video information. Another possibility for playing two video streams at the same time is to have one coming from CD-ROM and the other coming from hard disk. If multiple AVSS video processes are being run, care must be taken to provide adequate VRAM buffers for each process.

AVSS test tool program

The DVI Developer's Software Package contains a tool program which is very valuable for working with AVSS files: the **vplay** program, which uses AVSS to play .avs files for test purposes. This program works from command line parameters.

Summary

Motion video compression must be concerned with the speed of both compression and decompression. At the same time, greater compression must be achieved in order to fit into limited data rate storage devices. The technique of motion compensation is used to exploit the frame-to-frame redundancy in motion video signals.

DVI production-level high quality motion video compression uses a large computing facility to perform compression, while playback is on the DVI systems itself. DVI real time compression operates entirely on the DVI system and is useful during application design for experimenting with compressed video, and for applications which require real time compression.

AVSS (audio/video support system) is the DVI runtime software for playing motion video. While we refer to the i750 chip set as being the hardware heart of DVI technology, we should probably call AVSS the software heart of the technology. AVSS is the software which exercises the full audio/video/computer capabilities of the technology — it is what allows an application designer to mix and match any or all of DVI's presentation modes at the same time.

13

Video Manipulation — Special Effects

We touched on video manipulation briefly in Chapter 10 where some of the graphics functions that apply to still images were discussed. In this chapter we will take that subject much further and show a variety of things that can be done with still images, as well as discuss the possibilities for manipulating moving images.

Special Effects

In the television business, video manipulation is used to produce what are called *special effects* — pictures which have somehow been modified after they came from a camera in order to produce some desired impact on the viewer. In today's television broadcasting business, special effects have been brought to a high degree of sophistication. We will consider special effects in two classes:

- *transition* effects — these are effects used to enhance (or sometimes cover up) the transition from one camera shot to another. Common transition effects are fades, dissolves, wipes, etc.

- *running* effects — these are effects applied while a scene is taking place. Common types of running effects are overlays or mattes.

In fact, video effects got their start in the movie (film) business. Even early black-and-white movies made use of many types of special effects, which are similar to what we see today on television. In the film business, effects are created during the printing of a movie film — this is an optical process, and film special effects are often referred to as *opticals*.

When television broadcasting began, the number of special effects available was very limited relative to what film producers had become used to. This was because television was both analog and real-time, and doing analog video manipulation was difficult and expensive (it still is). However, early television engineers persevered, and many real-time transition effect devices were developed.

The first breakthrough in television effects came after video tape recorders were invented. Now television signals could be captured and stored just as with film, and special effects no longer had to be done in real time. This led to the introduction of many more effect types, which were based on video tape editing.

However, the real maturing of special effects for television came with the introduction of digital video technology to broadcasting in the early 1980s. Numerous digital effect-generating products are now on the market. Even though the television system itself is basically an analog system, digital effects are so much more desirable than analog effects that it is worthwhile to convert the analog television signal to digital, do an effect, and convert the signal back to analog to go through the rest of the system. The process of analog-to-digital conversion and digital-to-analog conversion is now essentially transparent to NTSC television signals.

The digital effect equipment used in broadcasting is extremely powerful — it can perform nearly any kind of transition effect that a producer could think of between a number of signals (not just two), and it also can do real-time running effects such as moving or warping motion video and then inserting that into another motion video scene. The kinds of effects shown on the network news programs are good examples of what can be produced today.

The broadcast digital effects systems are expensive, usually costing more than $100,000. What we will be talking about here is a capability to do some of the same kinds of effects in a PC-based system costing less than $10,000. However, we do not mean to imply that a PC-based system will compete exactly with the broadcast devices — you really do get what you pay for! Although the PC-based system does some of the same effects, it does them at lower resolution, sometimes slower than the systems which cost more than ten times as much. However, as PC digital video systems evolve, that performance gap will surely be narrowed.

Even in broadcasting, special effects are not always done in real time. Some broadcast video recorders have the capability to record video frames one at a time, and therefore certain effects can be done with a general-purpose computer that processes frames one at a time and then records each frame onto video tape as it comes out, however slow. Extremely elaborate picture manipulation or animations can be done this way. An entire industry has grown up using this technology to produce computer animations for both motion pictures and television. Therefore, another aspect of special effects is whether the effect is produced in real time or not. In the following discussion, we will cover only the effects done in real time with DVI Technology.

The subject of speed warrants more discussion. When talking about compression, we spoke of the number of processor instructions for each pixel. That is also a good way to evaluate the potential of a system for doing effects in real time. For example, if we are displaying 30 frames per second at a resolution of 512x480 pixels, we are talking about 7,372,800 pixels per second. With that number, a 12.5 MIPS processor like the 82750 PA can offer less than two instruction cycles per pixel. It is clear that the PA will not do very much manipulation at that resolution and speed. However, if we lower the resolution to 256x240 pixels, we have gained a factor of four, and now the PA can run almost seven instructions per pixel. With the power of the PA instruction set, significant real-time manipulation could be done at 256x240 resolution and 30 frames per second using seven instructions for each pixel.

However, if we also expect the PA to do decompression of motion video at the same time, we are back in trouble because full-screen decompression at 30 frames per second needs nearly all of the PA's cycles. Therefore, with the present speed of the 82750 PA we can only manipulate decompressed motion video at less than full-screen image size and/or with a lower frame rate than 30 frames per second. Intel has a long range plan to improve the PA's speed by introducing faster models. The first of these new devices will have a 2x speed increase; that one is nearing introduction as this is being written.

Still Image Transition Effects

There are many transition effects between still images which can be produced using only the **GrCopy( )** function. Because this function is so fast (3,000,000 16-bit pixels/second), it may be called repeatedly to produce an effect. A basic concept of doing this is the idea of first loading an image to an off-screen buffer in VRAM and then creating the transition by a series of repeated copies of parts of the image from the off-screen buffer to the bitmap that is being displayed.

We will now show some software examples of transition effects in DVI Technology. For the non–software reader, it will be awkward to summarize the points made at the end of all the examples. In this case, it is recommended that a non–software reader skip over the listings and the code descriptions, but read the explanations about how each effect works. This will give you the idea of how effects are created and what the possibilities are for generating effects that are different from the ones shown here.

Wipe transition effect

One of the simplest still-image transition effects obtainable with copying is the wipe. A horizontal wipe, for example, is done by repeatedly copying a 1-pixel-wide vertical strip of the image from the off-screen buffer to the screen, beginning at the left edge of the picture and moving progressively across to the right by 1 pixel as the copy repeats. We can demonstrate the wipe effect in the context of the demo program we started in Chapter 9, where we will do wipes with 16-bpp images.

Listing 13.1 is the code for the **do_wipe( )** function which can be added to demo.c. An item "7 WIPE" is added to the main menu of the program in order to call **do_wipe( )**. The function will wipe a different image onto the screen every time it is called. That is done by alternating between two different images.

```
/************** Start of Listing 13.1 ************/
/*  Load a different image and wipe it to the screen  */
void    do_wipe()
{
I16     x,width,height;

/*  Load a different image to BMw    */
x = load_image(&width,&height);
```

```
if (x < 0) return;

/*  Now wipe it to the screen    */
for (x = 0; x < width; x++)  {
   GrCopy(pBM, (SC_WIDTH-width)/2 + x, (SC_HEIGHT-height)/2,
          1, height, pBMw, x, 0);
     }
}
/************** End of Listing 13.1 **************/
```

As you can see, the code for a wipe is extremely simple. We first call **load_image( )** (explained in Chapter 10) to load a different image into the off-screen working bitmap **BMw**. Then we simply set up a loop with a **for( )** statement to repeat a copy as many times as there are pixels across the new image. The loop simply contains one **GrCopy( )** call, copying a 1-pixel vertical strip to the screen bitmap **BM** each time as the variable **x** is incremented across the image. The result is about a 1-second wipe from left to right across the screen. That is about the correct duration for a good subjective effect. If the speed needed to be faster, one would have to copy a wider strip each time, because most of the time in this routine is taken up by the overhead of the repeated calls to **GrCopy( )**, and by moving from one line to the next, copying only 1 pixel each time. Figure 13.1 (see color plates) shows some screen photos taken during the effect.

Cursor wipe transition effect
The wipe effect is easily modified to have a cursor of any color that moves across the screen during the wipe. This is done by adding a call to **GrRect( )** inside the **for( )** loop to draw the cursor ahead of the block which is being copied, as shown in Listing 13.2.

```
/************** Start of Listing 13.2 ************/
/*  Load a different image and wipe it to the screen
 *  using a cursor  */
void    do_cwipe()
{
I16     x,width,height;

/*  Load a different image to BMw   */
```

```
x = load_image(&width,&height);
if (x < 0) return;

/* Now wipe it to the screen    */
GrBmSet(pBM, GrBmDrawColor, colors[5]);      /* red */
for (x = 0; x < width - 2; x++)  {
   GrCopy(pBM, (SC_WIDTH-width)/2 + x, (SC_HEIGHT-height)/2,
          1, height, pBMw, x, 0);
   GrRect(pBM, (SC_WIDTH-width)/2 + x + 1, (SC_HEIGHT-height)/2,
          2, height);
   }
/* Now copy the last strip, covering up the cursor */
GrCopy(pBM, (SC_WIDTH-width)/2 + x, (SC_HEIGHT-height)/2,
          2, height, pBMw, x, 0);
}
/*************** End of Listing 13.2 ***************/
```

So as not to leave the cursor on the screen after the effect is completed, the main **for( )** loop must be shortened by the width of the cursor bar, and the remaining pixels of the new image are copied with one more call to **GrCopy( )** after the **for( )** loop is finished. This covers up the cursor with the last strip of the new image. In Listing 13.2 a 2-pixel width red cursor bar is used. Figure 13.2 (see color plates) shows a screen photo taken during the cursor wipe effect.

A more elaborate cursor could be obtained by using an image or a pattern for a cursor. This could be put at the right side of **BMw**, and instead of the **GrRect( )** call, an additional **GrCopy( )** could copy the cursor from its storage location in **BMw**. Very fancy effects can be done this way.

Iris transition effect

A transition which opens up the new image from the center is easily done with repeated **GrCopy( )** calls. We call this an *iris* effect because it is something like opening up a camera lens iris, although in this case the iris is rectangular. Listing 13.3 shows some code for doing this effect.

```
/*************** Start of Listing 13.3 ***************/
/* Load a different image and open it up on the screen  */
void    do_iris()
```

```
{
I16     x,h,width,height,xc,yc;

/*  Load a different image to BMw   */
x = load_image(&width,&height);
if (x < 0) return;

width &= 0xfffe;        /*  Make width an even value   */
height &= 0xfffe;       /*  Make height an even value  */

/*  Calculate the center starting point   */
xc = SC_WIDTH / 2;
yc = SC_HEIGHT / 2;

/*  Now iris the new image to the screen   */
for (x = 1; x < width / 2; x++)  {
    h = x;
    if (h > (height/2)) h = height/2;
    /*    top line      */
    GrCopy(pBM, xc - x, yc - h, 2 * x, 1,
            pBMw, width/2 - x, height/2 - h);
    /*    right side   */
    GrCopy(pBM, xc + x - 1, yc - h, 1, 2 * h,
            pBMw, width/2 + x - 1, height/2 - h);
    /*    bottom       */
    GrCopy(pBM, xc - x, yc + h, 2 * x, 1,
            pBMw, width/2 - x, height/2 + h);
    /*    left side    */
    GrCopy(pBM, xc - x, yc - h, 1, 2 * h,
            pBMw, width/2 - x, height/2 - h);
    }
}
/************** End of Listing 13.3 **************/
```

A **for( )** loop is set up to count up to a number which is half the width of the new image. Then there are four **GrCopy( )** calls inside of the loop to copy the new image in four 1-pixel lines at the top, right side, bottom, and left side. These copies move out over the new image until it is all copied. Two precautions have to be included: the height and width must be made even values, and the expansion of the image height must stop when the full height of the new image has been copied (the code assumes that the image is wider than it is high). Figure 13.3 (see color plates) shows screen photos taken during the iris effect.

Random-block transition effect

An interesting transition may be obtained by copying small blocks of the new image to the screen, using random locations in the new image. The effect repeats until all the blocks of the new image have been filled in. Listing 13.4 shows source code for this effect. The trickiest part of this effect is getting a random sequence of numbers which does not repeat any values until it has completed one full set of values. The algorithm used was provided by Michael Keith of Intel Princeton Operation — it is a common algorithm, but Mike found some magic numbers which provide the desired nonrepeating characteristic for values between 0 and 1023. Therefore, we must divide our image into 1024 blocks to use this algorithm.

After loading a new image into **BMw** with a call to **load_image( )**, we calculate the size of the blocks to use in variable bsize. Then we enter a loop which will execute 1024 times. The random algorithm is first called to get a random value r between 0 and 1023. The random algorithm skips the value 7, so that value is deliberately inserted in the last pass of the loop (**v == 1023**).

```
/************** Start of Listing 13.4 ************/
/*  Load a different image and use block transition   */
void    do_bt()
{
I16    r = 5,x,v,y,width,height,xc,yc,bsize,xsize,ysize;

/*  Load a different image to BMw    */
x = load_image(&width,&height);
if (x < 0) return;

/*  Calculate the image starting point    */
xc = (SC_WIDTH-width) / 2;
yc = (SC_HEIGHT-height) / 2;

/*  Calculate the size of a block (assume width > height)   */
```

```
bsize = width / 32;
if (width % 32) bsize++;

/*  Now bt the new image to the screen    */
for (v = 0; v < 1024; v++)  {
    xsize = bsize;
    ysize = bsize;

    /*  Get a non-repeating random value between 0 and 1023  */
    if (r & 0x200) r = (r << 1) & 0x3ff;
    else r = (r << 1) ^ 9;
    if (v == 1023) r = 7;   /* insert the missing value last */

    /*  Calculate x and y coordinates from the random value  */
    x = (r % 32) * bsize;
    y = (r / 32) * bsize;
    if (y > height) y = height - 1;

    /*  Now make sure block will not go outside the image  */
    if ((x + xsize) > width) xsize = width - x;
    if ((y + ysize) > height) ysize = height - y;

    /*  OK, now copy the block  */
    GrCopy(pBM, xc + x, yc + y, xsize, ysize,
            pBMw, x, y);
    }
}
/************** End of Listing 13.4 **************/
```

Then the random value is partitioned into x and y coordinates by dividing by 32 and multiplying by the block size to get the x coordinate of the block, and then using the modulo (remainder) C operator with 32 to get the y coordinate for the block to be copied. The resulting block has to be tested to see if it extends outside of the image to be copied — if so, the block sizes **xsize** and **ysize** are reduced to fit. Finally we are able to copy that block to the screen. The result is a transition effect which takes about two seconds to compete — parts of the new image are appearing at random locations on the screen until the entire image is there. Figure

13.4 (see color plates) shows screen photos taken during the block transition effect.

All of the preceding effects have been done with only one of the DVI graphics functions — **GrCopy( )**. In fact, there are many other effects which can be done with this function. We have only shown effects so far which are practical at 512x480 resolution — many more things can be done if we reduce the resolution to 256x240. Let's look at a transition effect that is practical at 256x240 resolution.

Slide-on transition effect

A transition that is often used in television is the *slide-on* effect, where a new image moves across the old image or where the new image pushes the old image off the screen. Either of these require repeatedly copying a large block, up to full-screen in size — the large number of pixels involved in each copy makes these effects not practical at 512x480 resolution — at present 82750 PA copy speeds, they are just too slow. (An effective transition of this type must complete in 4 or 5 seconds, and the effect can only move by 1 or 2 pixels per step to maintain smooth motion — the PA is about four times too slow for this.)

Another consideration arises when doing repeated copies to a large part of the screen that is being displayed — the vertical scanning of the display and the repeated copying can interfere with one another and give a very uneven effect. If the copying time is short enough, it is possible to synchronize the copying to the vertical scanning by calling a microcode function which waits for a particular scanning line number. However, if we are copying full-screen as needed for the slide-on effect, there is no proper place to do the copies except the vertical blanking time. That is about 1 millisecond, whereas a full-screen copy at 256x240 resolution with 16 bpp takes about 20 milliseconds, so we are pretty far off. The solution is to set up two screen buffers and alternate between them, doing the copying on the buffer that is not currently being displayed. The 82750 DA can be switched between the buffers by calling **GrBmDisplay( )** appropriately. This will cause a transition between the two buffers during the next vertical blanking interval. The following code example, Listing 13.5, shows how that is done. It also introduces several other interesting DVI software functions.

```
/*************** Start of Listing 13.5 *************/
/* Switch to 256x240 res and do a slide-on transition effect */
```

```
void      do_slide()
{
I16       s = 0,x,height,width;
hGrBm     pBM256[2];

clearBM();

/* Initialize 256x240 screen bitmaps */
GrBmAlloc(&pBM256[0]);
GrBmAlloc(&pBM256[1]);
GrBmSetPacked(pBM256[0], GR_BM_16, 0x10000L, 256, 240, 256);
GrBmSetPacked(pBM256[1], GR_BM_16, 0x30000L, 256, 240, 256);

/* Switch to displaying 256x240 bitmap [0] */
GrBmDisplay(pBM256[0]);

/* Clear the keyboard buffer */
while (kbhit()) getch();

RtxPrintf("Press any key to stop the transitions\n");

while (!kbhit()) {
   /* Load a different image to BMw   */
   x = load_image(&width,&height);
   if (x < 0) return;

   /* Copy the screen to BMw right side */
   GrCopy(pBMw, 256, 0, 256, 240, pBM256[s], 0, 0);

   /*  wait until that is done */
   GrWaitDone( 200L );

   /* Now do the transition */
   for (x = 0; x < 257; x+= 1) {

       /* Copy new image from BMw to BM256[s ^ 1] */
       GrCopy(pBM256[s ^ 1], 0, 0, 256, 220,
               pBMw, 256 - x, 0);

       /* Wait until that is done */
```

```
        GrWaitDone ( 200L ) ;

        /*  Switch buffers  */
        s ^= 1;
        GrBmDisplay (pBM256 [s] ) ;
        GrBmDisplayWait (0) ;

        /*  Quit if a key is pressed  */
        if (kbhit ()) break;
        }
    }
/*  Free the BM256 bitmaps  */
GrBmFree (pBM256 [0] ) ;
GrBmFree (pBM256 [1] ) ;

/*  Return to 512 mode  */
clearBM () ;
GrBmDisplay (pBM) ;
menu_drawn = 0;  /*  flag to re-draw the menu  */
}
/*************** End of Listing 13.5 ***************/
```

In order to include this demo in the same program as the rest of our
demos, we will have to switch the screen resolution when we enter this
particular demo. The first part of the code in Listing 13.5 does just that.
In addition to setting up the usual height, width, and count (x) variables,
we will set up two bitmaps for the 256x240 screen buffers and a variable
s to keep track of what screen we are displaying. Next, we allocate the
two bitmaps **pBM256[0]** and **pBM256[1]** with two calls to **GrBmAlloc()**
and then fill in their formats with calls to **GrBmSetPacked( )**.

We will set up this demo so that it repeats continuously until a key
is pressed. Therefore, we should make sure that the keyboard buffer is
cleared before proceeding. Then we will put a prompt on the command
monitor screen to tell the user how to stop the demonstration.

Next we enter a loop which repeats until the C function **kbhit( )**
reports that a key has been pressed. The first operation is to call
load_image( ) to load a different image into **BMw**, the working buffer.
Note that this image is one of the same ones that we use in the 512x480
demonstrations, they are both larger than 256x240, so we will be

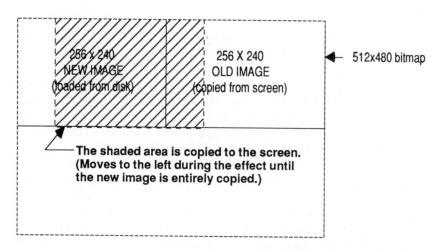

Figure 13.5   Layout of off-screen bitmap showing how images are copied to the screen to make the slide-on transition

showing only part of them on the 256x240 screen that we now have. Therefore, we can copy the current screen to the right half of **BMw**, and create an image in **BMw** which has the new image on the left and the old image on the right.

The slide-on transition is created by copying repeatedly from **BMw**, starting with the new image part of **BMw** and moving the source for the copy progressively to the left in **BMw**, exposing more and more of the new image coming in at the left as seen on the screen. This is diagrammed in Figure 13.5. The transition loop thus is simply a repeated single copy to produce the effect. However, to maintain proper synchronization with the scanning, we must do some other things in the copying loop.

Since each 256x240 copy takes about 20 milliseconds, we can set up to synchronize with scanning frames — doing one copy and switching buffers once for each frame. That will happen automatically if we simply wait until **GrCopy( )** finishes each time by calling **GrWaitDone( )** and then calling **GrBmDisplay( )** to show the newscreen. That works because **GrBmDisplay( )** will automatically wait until the next vertical blanking interval before showing the new screen.

The copy loop uses the counter **x**, which will go from 0 to 256 in steps of 1. The first step inside the loop is to call **GrCopy( )**, copying from **BMw** to the bitmap **pBM256[s ^ 1]**, which is the screen bitmap *not* currently being displayed (**pBM256[s]** is always the one displayed). Note that we are copying between bitmaps of different sizes — **BMw** is 512x480 and **BM256[s ^ 1]** is 256x240. That is fine with **GrCopy( )** — it can copy between two bitmaps of different sizes, as long as they have the same pixel format. Then we call **GrWaitDone( )** to wait until the copy is completed, at which point we can switch buffers by calling **GrBmDisplay(s)** after we exclusive-or'd **s** with 1 to make it toggle between 0 and 1. A final step is to wait until the new screen is actually displayed before beginning processing of the next screen. This is easily done with a call to **GrBmDisplayWait(0)**.

The resulting transition moves smoothly across the screen, moving at 1 pixel per frame. Figure 13.6 (see color plates) shows some screen photos of the effect. One pixel per frame means that it will take 256 frames to do the full screen, or about 8.5 seconds. This is more time than we said such a transition should take — it can be speeded up 2x by going 2 pixels per frame. However, close observation of the transition will show that the edge has a flickering gear-tooth pattern to it as it moves across the screen. This is a normal artifact caused by interlaced scanning of objects which move horizontally across a television screen. (Because of that problem it would not be satisfactory to move this transition even faster by going more than 2 pixels per frame — the gear-teeth get wider.) This effect would not occur if noninterlaced scanning is used.

A more complex approach is possible to reduce the gear-toothing artifact. Since the interlaced scanning is only displaying half of the lines in each vertical scan, we could arrange to only copy half of the lines during the time of one field — half of the lines contain half as many total pixels, so such a copy will fit in one field time. Then we can switch the display at field rate. (The lines for one field can be extracted by defining a temporary bitmap at the same address as **BMw**, but it has twice the pitch value of **BMw**. Copying 256-pixel lines from this new bitmap will extract every other line.) Now, by moving the copy by one pixel every field instead of two pixels every frame, the effect will have the same duration, but it will be even smoother.

When the transition is completed, the keyboard is tested for a key pressed. If none, a new image is loaded and another transition is started. When a key does get pressed, we free up the bitmap memory space by calls to **GrBmFree( )**, and then clear the 512 screen bitmap **pBM** and switch back to it with a call to **GrBmDisplay(pBM)**. Finally, we zero the variable **menu_drawn**, which will cause the **menu( )** routine to redraw the menu bar.

The previous code example is one of the most complex ones we are going to do here. In addition to the techniques for the slide-on/push-off transition effect, the example showed how to switch resolution modes, how to use two screen bitmaps and switch between them on a frame basis, and how to synchronize that action with the display scanning.

Image Processing

Operations which change the values of pixels instead of simply moving them around are called *image processing* operations. The 82750 PA can be used effectively to do many image processing tasks. In Chapter 10 and Figure 10.5, we showed still-image examples of a variety of things which can be done. Some of these functions, such as **GrScale( )**, are too slow for real-time effects unless the image is extremely small, but others can be used to create interesting effects in real time.

For example, **GrPolyWarp( )** can be used repeatedly to zoom an object onto the screen. The speed of **GrPolyWarp( )** is about 500,000 pixels per second, so a warp of a 128x128 object could be done approximately 30 times per second. (Only up to 1:1 size, however — the speed is calculated based on the *destination* pixels.) Effects such as continuously rotating a small object are very practical and effective with **GrPolyWarp( )**.

In Chapter 10, we saw a list of the image processing functions. Some possible uses of these functions for special effects are

1. The function **GrBrightness( )** can be used to highlight or to dim a portion of the screen by raising or lowering the luminance values in a rectangle to produce a spotlight kind of effect. Similarly, the **GrContrast( )** function can be used for the same purpose. Depending on the nature of the images being highlighted or dimmed, one or the other function will be most effective.

2. The **GrBlend( )** function can do a simple fade. The two images to be faded are loaded off-screen and then **GrBlend( )** is called repeatedly to fade from one to the other on the screen. This is very effective with 256 x 240 images, it is a little slow for 512 x 480 images.

3. The **GrMosaic( )** function can be used to create a transition where an image fades out by going to progressively larger and

larger pixels until only one pixel is left. The reverse is also possible for bringing an image in by growing it out of one very large pixel.

The 9-bpp Mode

The previous examples of transition effects were all done with 16-bpp mode. As was explained, the scope of effects possible is limited by the speed of the 82750 PA in manipulating pixels, and we found that more effects were possible at lower resolution (fewer pixels). Further improvement can be obtained by going to smaller pixels — and a good step in that direction is to use the special 9-bpp mode, which uses 8-bit Y with chrominance subsampled by 4:1 in each direction. This yields a nearly 2-to-1 improvement in speed for pixel operations.

The DVI graphics copy and drawing primitive functions will automatically handle the details of drawing into 9-bpp bitmaps. The only difference from working with 16-bpp mode is to make sure that the drawing functions are given the correct color values for 9-bpp by calling the **GrPixFromColor( )** function to get the color value to use. Once this is done, a single call to any graphics function will draw correctly into both luminance and chrominance bitmaps. However, it should be recognized that color graphics drawn into subsampled chrominance bitmaps will be interpolated up to full resolution by the 82750 DA when displayed, and this will sometimes cause color fringing effects. To avoid this with graphics drawn on subsampled-UV bitmaps, one has to use one of the optional mixed video/CLUT modes available in the DA.

Use of 9-bpp format for effects

Because the 9-bpp format is used for all motion video and for the lossy still compressed images, we will cover it in some more detail. We will do a wipe effect that creates a four-way mirror image of a picture. Listing 13.6 shows the source code for this effect.

```
/************** Start of Listing 13.6 ************/
/*  Switch to 512x480 9Y mode and do a wipe   */
void    do_9bpp()
{
I16     x;
hGrBm   pBMs,pBMws;
```

```
GrPix   color;
/* Set up two 512x480 bitmaps in 9Y mode */
GrBmAlloc(&pBMs);
GrBmAlloc(&pBMws);
GrBmSetPlanar(pBMs, GR_BM_9Y, 0x90000L, 512, 481, 512, 0xd0000);
GrBmSetPlanar(pBMws,GR_BM_9Y, 0x10000L, 512, 481, 512, 0x50000);

/* Create a color look-up table for pBMs:
 * white,yellow,cyan,green,magenta,red,blue,black,gray,
 * lightblue */
GrBmSetIndexRgb(pBMs, GrBmVidClut16, 0, 255, 255, 255);
GrBmSetIndexRgb(pBMs, GrBmVidClut16, 1, 255, 255,   0);
GrBmSetIndexRgb(pBMs, GrBmVidClut16, 2,   0, 255, 255);
GrBmSetIndexRgb(pBMs, GrBmVidClut16, 3,   0, 255,   0);
GrBmSetIndexRgb(pBMs, GrBmVidClut16, 4, 255,   0, 255);
GrBmSetIndexRgb(pBMs, GrBmVidClut16, 5, 255,   0,   0);
GrBmSetIndexRgb(pBMs, GrBmVidClut16, 6,   0,   0, 255);
GrBmSetIndexRgb(pBMs, GrBmVidClut16, 7,   0,   0,   0);
GrBmSetIndexRgb(pBMs, GrBmVidClut16, 8, 100, 100, 100);
GrBmSetIndexRgb(pBMs, GrBmVidClut16, 9, 100, 100, 255);

/* Clear the new screen bitmap BMs to gray */
GrBmSet(pBMs, GrBmDrawOutline, FALSE);
GrPixFromColor(pBMs, GR_COLOR_RGB, 100, 100, 100, &color);
/* Make sure it is a video color */
color |= 0x1L;
GrBmSet(pBMs, GrBmDrawColor, color);
GrRect(pBMs, 0, 0, 512, 481);
GrWaitDone( 200L );

/* Switch to display 512x480 BMs with mixed video/graphics */
GrBmDisplay(pBMs);

/* Load an image into BMws   */
x = load_9bpp(pBMws, "samp256s");
if (x < 0) {
      printf("Image did not load\n");
       goto end;
       }
```

```
/* wipe it to the screen  four ways with mirrors  */
for (x = 4; x < 256; x++) {
    GrCopy(pBMs,         x, 4, 1, 237, pBMws, x - 4, 1);
    GrCopy(pBMs, 508 - x, 4, 1, 237, pBMws, x - 4, 1);
    }
for (x = 0; x <  237; x++)
    GrCopy(pBMs, 0,  240 + x, 512, 1, pBMs, 0, 240 - x);

/* Create a prompt box with CLUT mode  */
/* First, set to draw only to Y-plane  */
GrBmSet(pBMs, GrBmDrawPlane, GR_PLANE_Y);

/* Now write clut index values into Y only  */
GrBmSet(pBMs, GrBmDrawColor, 0x0);  /* CLUT white  */
GrRect(pBMs, 356, 216, 140, 48);
GrBmSet(pBMs, GrBmDrawColor, 0x98);  /* CLUT ltblue */
GrRect(pBMs, 358, 218, 136, 44);
GrBmSet(pBMs, GrBmDrawColor, 0x0);  /* CLUT white  */
GrText(pBMs, 364, 224, "PRESS ANY KEY");
GrText(pBMs, 384, 244, "FOR MENU");

/* wait for a keystroke  */
while(!kbhit()) ;
/* clear keyboard buffer  */
while (kbhit()) getch();

end:
/* Free the subsampled bitmaps  */
GrBmFree(pBMs);
GrBmFree(pBMws);

/* Return to 512 mode  */
clearBM();
GrBmDisplay(pBM);
menu_drawn = 0;  /* flag to re-draw the menu  */
}
/************** End of Listing 13.6 **************/
```

This function will operate with a 512x480 Y, chroma-subsampled display using the mixed video/graphics mode. Control of whether a particular area of the display is video or graphics is by means of the least significant bit of the Y value. If the Y least significant bit (LSB) of a pixel is a 1, then the pixel will be in video mode (Y, VU subsampled). If the Y LSB is a zero, then the pixel will be in color look-up graphics mode, using the upper 7 bits of the Y value as an index into a 128-entry color look-up table. We will explain how to set that up in a moment. (There is also a mixed video/graphics mode where the mode switching is controlled by the LSB of the U value. This allows indexing a 256-entry color look-up table, the trade-off is that mode switching control is only on a 4x4 block of pixels at a time.)

The function begins with the usual declaration of local working variables. We declare two bitmaps: one for the screen and one for off-screen. We also declare a 16-element array of type **U32**. The next step is to get memory space for our two bitmap descriptors using **GrBmAlloc( )**. Then we call **GrBmSetPlanar( )** to initialize the values for these bitmaps. We will put our screen bitmap **BMs** at page 9 in VRAM (0x90000) so that it does not overlap the 16-bpp screen bitmap at VRAM address 0x10000. This way, we can set either of these bitmaps up before switching to it and thus get a good transition between modes. (If you display in 9-bpp mode an area of VRAM which has 16-bpp data in it, you will get a very garbled display.) Since a 512x480x8-bpp bitmap takes less than four pages of VRAM, there is plenty of room above page 9 for this bitmap at full size.

The Y, chroma-subsampled format requires two bitmaps: one for the luminance, as we have just discussed, and another smaller bitmap for the chrominance. The last parameter of **GrBmSetPlanar( )** defines the location of the chrominance bitmap. It will be subsampled by a ratio of 4 to 1 in each direction and it will be located just above the Y bitmap at page 12 (0xd0000).

The second bitmap **BMws** is set up to be at address 0x10000 in VRAM, which overlaps the 512x480x16-bpp screen bitmap. However, we will not use this in 9-bpp mode until we have switched the screen away from this area of VRAM. The rest of the parameters of **GrBmSetPlanar( )** are the same as the screen bitmap.

Since we are going to use the mixed video/graphics mode in this example, we must have a color look-up table. These tables can be stored on disk in image files, either by themselves, or with images, however in this example we will make our own. The look-up table is a 16-element array of type **U32**, which we have already declared. Values can be generated for that table from R, G, B parameters with the call **GrBmSet-IndexRgb( )**. This call generates the physical color table format required by the 82750 DA. We have created a table of 10 colors containing the color bar colors and a few others. Six more could be added to fill up the table, but we don't need them for this example.

Now we are ready to clear the new screen bitmap. We will make it gray. We get the color value by calling **GrPixFromColor( )** with RGB values for gray, and telling that routine about **BMs**. (We have to do this because the 9-bpp color value is different from the 16-bpp color value we put earlier in the array **colors[ ]**.) Once getting the color from **GrPixFromColor( )**, we also have to set the Y LSB to put the entire screen in video mode. This is done by OR-ing the color value with 0x1. (The format of a color value for 9-bpp mode is 8 bits each of Y, V, U with Y least significant, thus the least-significant bit of Y has a hex value of 1.) Then we have to place this color value in DrawColor by using a **GrBmSet( )** call. Finally we can call **GrRect( )** to do the clear. This function will clear both bitmaps of **BMs** in one operation.

Since **BMs** is now ready, we can switch to displaying it by calling **GrBmDisplay( )**. This function loads the CLUT into the 82750 DA before it switches the display — it all happens during one vertical blanking interval.

Now that we are displaying **BMs**, we can load an image off-screen into **BMws**. We will use the image **samp256s.***, which is one of the test images which comes with the DVI Development Software. It is 256x256 in 9-bpp mode and consists of three files in .imy, .imi, .imq format. The function **load_9bpp( )** takes care of this for us — we will look at that function in Listing 13.7.

The effect we produce is to do a double wipe starting at the left and right sides of the screen, bringing in the image on the left and its mirror image on the right. When this has completed in the top half of the screen, we do a second vertical wipe to create a mirror image of the top half of the screen in the bottom half of the screen. The mirroring is obtained by copying 1-pixel strips in reverse order. The entire effect takes about 6 seconds and the result is shown in Figure 13.7 (see color plates).

An important concern in working in 9-bpp mode is to keep the luminance and chrominance values properly together in space, considering that the chrominance is subsampled by 4:1 in each direction. This is usually taken care of by using pixel coordinates that are multiples of four, or in the case of loops, starting from a location whose coordinates are divisible by four. If this is not done, color fringes may appear because the luminance and the color become misregistered.

At the end of this effect we will place a prompt box on the screen to tell the user to press a key to continue — we will do this in CLUT mode by locally switching the Y LSBs where the prompt box will appear. In order to achieve a smooth transition from video mode to CLUT mode, the chrominance values should be preserved around the prompt box as much as possible. This can be achieved by only switching the Y values, and leaving the U and V planes untouched. It is accomplished by calling **GrBmSet( )** to set GrBmDrawPlane to GR_PLANE_Y. Then we can draw the prompt box using the normal graphics and text calls with GrBmDrawColor set to the appropriate CLUT index values for the colors we want.

Indexing into a 16-element lookup table is only going to take 4 bits of data, but the table indexing is arranged in the 82750 DA so that it is done twice — once for luminance and once for chrominance. This gives additional flexibility in that the luminance and chrominance values can be separately indexed from different entries of the 16-element table. Effectively, one can get 128 colors this way — 8 luminance values combined with 16 chrominance values. (Note that only even luminance values can be used in CLUT mode — odd values have the LSB set, which will switch the pixel to video mode.) However, in a practical situation, some of the CLUT combinations of luminance and chrominance will not be useful — for example, the luminance from yellow combined with the chrominance value from blue will not make a brighter blue because the blue by itself was already as bright as it can get. However, the system will reproduce all the combinations — it's just that some of them will not look like what you might have expected.

The color lookup table index value has the Y index in the four least significant bits and the chrominance index in the next higher four bits. The remaining bits above that are not used when we are drawing to the Y-plane only. Therefore, an index of 0x98 will get light blue from our table, while 0x96 would get light blue color with luminance from the blue table entry.

The screen is now completed, and we wait for a user keystroke. When that happens, we clear the keystroke buffer and then free up the bitmap structures we had been using and switch back to 16-bpp mode with a call to **GrBmDisplay( )** pointed to bitmap **BM**.

Now let's look at the function for loading a 9-bpp image, shown in Listing 13.7.

```
/************** Start of Listing 13.7 *************/
/* Load 9-bit image files into a bitmap */
I16    load_9bpp(pBMss, filename)
hGrBm  pBMss;
char   *filename;
{
I16    x,width,height,xlen,ylen;
char   name[64];
hGrImHdr ImHdr;

GrImHdrAlloc(&ImHdr);

GrBmGet(pBMss, GrBmXLen, &xlen);
GrBmGet(pBMss, GrBmYLen, &ylen);

/* Look for the file and get some data about it       */
GrImHdrAlloc(&ImHdr);
strcpy(name,filename);
strcat(name,".imy");
x = GrImLoadHdr(name, NULL, pBMss, ImHdr);
if (x < 0) goto end;
GrImHdrGet(ImHdr, GrImHdrXLen, &width);
GrImHdrGet(ImHdr, GrImHdrYLen, &height);
GrImHdrFree(ImHdr);

/* Load the image   */
x = GrImLoad(filename, NULL, pBMss, 0, 0, width, height);
if (x < 0) goto end;

x = 0;
end:
GrImHdrFree(ImHdr);
return(x);
}
/************** End of Listing 13.7 **************/
```

This function **load_9bpp( )** takes a bitmap handle and a filename and performs loading of .imy, .imi, and .imq files. The function begins by setting up local variables for the normal housekeeping.

We use the same procedure of getting the image file header (from the .imy file) that we used in the 16-bit mode to find out the pixel size of the image, which we will need later. To load the parts of the image, we use **GrImLoad( )** as we did for 16-bpp images; that function can tell from the type of the bitmap that we specified to load all three color component files.

However, since we are using the mixed video/graphics mode for display of this file, we must be sure that the image files being loaded have the Y LSB properly set for video mode (assuming the entire image is video, as in a digitized photograph). This could be done by OR-ing the pixels with 0x1, using **GrRect( )** with GrBmRop2 set to GR_ROP2_OR by calling **GrBmSet( )**. (Note that we should put Rop2 back to GR_ROP2_NONE when we are finished so that subsequent functions will work properly.)

The previous example shows the general principles for using the 9-bpp chroma-subsampled mode. This mode gives nearly a 3:1 data compression compared to 24-bpp images as well as significant speed advantages for image retrieval and image manipulation. It is well worth the small additional complexity of using it. Further, if your application expects to mix graphics and motion video, it will be best to do everything in 9-bpp mode because that is the mode required for motion video.

Effects with Motion Video

In Chapter 12 we spoke of doing graphics operations while motion video is playing. One possibility of course would be to do some copy or warp operations to manipulate the 30-frames-per-second motion video. The AVSS software provides for this with the hook routine concept. A HookAfterDecode routine could be written, for example, to cause a motion video window to be wiped onto an existing screen. This would be done by having the existing background screen in a separate bitmap, and at HookOnVbi time (just before display of a motion frame), part of the background screen would be copied to cover up the motion frame. As frames progressed, the size of the background copy would be reduced until it was all gone — this effectively would uncover the already-moving motion image.

The catch of doing this is of course the speed of the 82750 PA. A full-screen motion image uses up all of the PA's cycles for decompression, and we have seen that a full-screen (256x240) copy also takes all of the PA at

30 frames per second. Therefore, to do both of these things together, we either have to reduce the number of pixels in the motion window or we have to reduce the frame rate. There are good possibilities in either of these directions — the choice would be determined by the needs of a particular application. The point here is simply that some compromise is required to do effects with motion video.

The idea of using **GrPolyWarp( )** with motion video is a bit more difficult to accomplish, because **GrPolyWarp( )** is about five or six times slower than **GrCopy( )**. It should be reasonable to warp a motion image containing about 10,000 pixels and still maintain 30 frames per second. That is only about 1/6th of a 256x240 screen. However, some applications may want to do that. Note that there is no restriction to having any kind of effect contained in the analog motion video that gets compressed. So, if you want a full-screen zoom effect at the start of your motion video, have that done in video postproduction before you send your video for DVI presentation-level compression. (Zoom, or other effect capabilities may eventually be added to the PLV compression process as options— that would save the cost and performance degradation of postproduction effects.) You'll then get your effect by normal playback of the compressed video.

An effect which is often used with motion video is the fade-out. This effect can be produced by the **GrBlend( )**, as a still-image effect. It is set up to be executed beginning immediately when the last frame of a motion sequence is displayed. If the fade-out is fairly fast (2 seconds), most viewers do not even notice that the motion has stopped during the fade.

Summary

Video special effects can add a lot to a presentation — from our experience with television productions, we expect smooth transitions and interesting variations in the way visual material is displayed. A PC-based video system has the capability to do almost any kind of effect. However, the practical consideration of processing speed limits what actually is effective. The addition of a fast, programmable video-processing engine like the 82750 PA in DVI Technology greatly expands the possibilities for doing real-time video effects. Even so, today this is largely restricted to effects on still images because video decompression of motion images already uses up the fast processor.   In this chapter we saw programming examples of effects like wipes, block transitions, and slide-ons. These were all done with the DVI graphics primitives for copying images. It is also important to note that as video processors become faster in the future, fancier effects will become practical, both for still and motion video.

14

Developing a DVI Application

The task of designing an application for an interactive PC-based audio/
video system is very complex and involves many different disciplines. In
this chapter we will review the steps of planning and designing an
application of DVI Technology. We will choose a simple example and
plan it out. Then we will look in more detail at some of the programming
for that application, considering both C-language programming or the
use of authoring systems.

Because of the large amount of work required for most interactive
audio/video/computer applications, and the wide diversity of skills
required, an application design project usually involves a number of
people — often 10 or more. Not all these people will be full-time, but they
all must contribute their special skills to the project. Therefore, management of the project becomes a non-trivial task, and the first thing to do
is to appoint someone to be project manager. The project manager job
itself is not technical, and it will not necessarily be a fulltime task over
the duration of the project, so the project manager may have other re-
sponsibilities. However, the first criterion for a project manager is his/
her breadth of background and managerial skills. The project manager

will be responsible for accomplishing the planning steps outlined below.

As in any design project, one begins application design by identifying:

1. What is the goal of the application?
2. Who is going to use the application?

These questions are so simple that they may not seem important, but they are fundamental to finding the direction for an application design project. There has to be an understanding about what the objective of the project is and who is going to be the customer (end user) for the product. In the discussion that follows, we will begin building answers to these and other questions for a hypothetical application which we will use as an example. It will not be possible to present a complete design in one chapter, but this example will give us a context in which to bring together all of the elements of an application design project.

A Sample Application

For our example we will choose an application which does not require you (the reader) to have any additional subject-matter knowledge about what is needed. Therefore, our example application will be to make a version of this book as an audio/video/computer application. This application could be made available to users of DVI Technology as a training tool and as a resource of information. Using CD-ROM as the medium, we can in fact go much further than one book and can include additional levels of detail in many areas. For example, the entire software documentation for DVI Technology can be easily included. We'll call the application "DVI Technology Data Base," and let's try to answer the two questions we posed above. We'll answer each one with a paragraph:

1. What is the goal of the application?

   This will be an educational application for developers of audio/ video/computer systems. It will teach all the aspects of setting up, using, and programming for such systems, specifically with DVI Technology. Using CD-ROM, the application will make available the detailed information about systems at several levels, all the way down to the detailed hardware and software documentation. The application must make it easy for the user to explore information at any level.

2.   Who is going to use the application?

The users of the application will be technical people who need to know about audio / video / computer systems. The users already know how to operate a personal computer.

The above application concept may seem trivial, considering that all the information included is already present in books like this one and in system documentation manuals. However, such information can be made much more effective and accessible by the use of audio/video presentation techniques.

Knowing what we want our application to do still leaves us a long way from being able to start any actual design work. Now we have to think about *how* the application is going to do its thing. An approach to this step is to write what video producers call a *treatment*. This is simply a narrative (typically a few pages) which describes in words how the application should work and how it should look. The treatment is something which anyone should be able to understand (it should not be technical), and it is a vehicle to get everyone on the project into agreement about what the product will be. Usually a single person will draft a treatment, and then the entire project team will review and critique it and make revisions until a document is generated that everyone agrees with. This process could take some time, particularly with a group that has not worked together before. Whatever time and effort it takes is well worth it to eliminate as many disagreements as possible before any great amount of work has been done.

Therefore, here's a treatment for our sample application:

DVI Technology Data Base
Treatment

This application is intended for developers of application programs using DVI Technology. It contains on one floppy disk and one CD-ROM disc, all the hardware and software documentation, demonstration programs, and other writings about DVI Technology. The users of this program are technically sophisticated in at least one of the aspects of video / audio /

computer application development.

The application begins with an opening high-resolution video still, created by a graphic artist, which contains the application name, copyright notices,etc. At the top of the screen is a menu bar. A cursor appears for use in making selections. An appropriate menu bar will always be at the top of the screen unless the task being run requires the full screen. (Full-screen motion video, for example, is by definition going to require the full screen!)

The menu provides top-level access to the following things:

- *A video sequence introducing DVI Technology*
- *A video and graphics presentation about installation of DVI hardware and software*
- *DVI demonstration programs*
- *The complete text and illustrations of this book*
- *The hardware and software documentation for DVI Technology accessed via an index*
- *DVI programming code examples*

Text-based materials listed above will be displayed in a window with complete support for easy readability. As subjects appear for which there is additional information anywhere else in the application, they will be highlighted to indicate that more information is available, and the user may view the additional information by placing the cursor on the highlight and selecting. Figures and drawings which go with the text will also be displayed in a scrollable window.

DVI demonstration programs may also be called from this application. This should be set up so that return will be made to this application when any demonstration completes. All the software required for this program and all of its sub-programs (demos, for example) should be contained on the two disks of this program. The floppy-disk part of the program should also be capable of being copied onto a hard disk and run from there. However, no other use of a hard disk should be required.

The above treatment may seem fairly simple, but in fact it has already indicated a lot of work which must go into the project. The next step is to consider what kinds of tasks will be needed to implement an application following the concepts in the treatment. We'll make a list of the tasks which are required, separating programming tasks and non-programming tasks:

Nonprogramming Tasks	Programming Tasks
Create screen design and art Digitize the artwork	Program the screens
Opening screen/menu	Program the menu
Design data display windows Program display of illustrations	Program text scrolling
Produce the video introduction Compress it	Playing of intro
Produce the installation video Compress it	Playing of installation video
Acquire the data, select exact material to include Prepare the index	Program for indexing
Acquire demonstration programs	Program to call them and return to this application
Plan exact contents of CD-ROM	
Set up the main program	
Data conversion and digitizing	
Initial simulation and testing	Fix bugs and update
End-user testing	Fix bugs and update
Master the CD-ROM	
Final testing	Fix bugs and update (floppy disk contents only)

This chart shows that there are many tasks in addition to the programming of the application. But there's still more planning to do — we cannot do any artwork, production, or programming until the above list has been reduced to a set of functional specifications. This means that most of the nonprogramming tasks need to have some preliminary investigation in order to write the specifications. For the proposed application here, that would particularly apply to the data and its handling. The available data must be collected, reviewed, and sized in order to plan the application's contents. Until the exact data types to be used have been determined, it is impossible to plan the kinds of programming that will be required to index it and to display it.

Also, there are key decisions to be made, such as how the user interface will be implemented. This could be based on mouse, joystick, keyboard, or combinations.

Each of the video/audio production tasks needs a treatment of its own to begin the planning of contents. Then scripting and planning for production can begin.

Data planning

A key step at this point in the development process is the planning of the data requirements and how the data is going to be formatted and displayed. Below is a list of the data types we would like to consider for this application and some notes about formatting and sizing. (The numbers given are approximations for the purpose of this example.)

Analysis of data needs

Data type: DVI software documentation
Present format: VAX text files — 1 megabyte

Data type: DVI hardware documentation
Present format: Text: VAX text files — 0.5 megabyte
                 Line drawings: CAD vector format — 50 8 in. x10 in. pages

Data type: C code examples
Present format: MS-DOS ASCII files — 1 megabyte

Data type: Book — "Digital Video in the PC Environment"
Present format: Text: Microsoft Word (Macintosh) — 0.7 megabyte
                 Line drawings: MacDraw format — 25 8 in. x10 in. pages
                 Photographs: film negatives (50 total, most are shots of the DVI screen)

Data type: DVI demo programs
Present format: Programs and image files: MS-DOS format — 10 megabytes

Motion video and audio: AVSS files (MS-DOS) —
50 megabytes

Data type: Motion video
Video introduction (full screen) — 2 minutes
Installation video (192x160 window) — 5 minutes

Data type: Index files
Present format: To be developed. Estimated size — 1 megabyte

This totals about 64 megabytes of existing digital data, a lot of drawings and photographs in other formats which will have to be digitized, and motion video which will go to presentation-level compression. In order to see where we stand on total data, we will have to figure out how the drawings and photographs will be handled.

Handling line drawings

The line drawings for the hardware documentation and for this book are in vector or object-oriented formats. DVI Technology does not have software to directly display these formats. The simplest approach is to print hard copy from the systems that store the drawings and digitize the hard copy using a high-resolution scanner. They can then be displayed as DVI images. (Going to hard copy in this computer age seems backward, but for the small number of drawings involved here it is more economical than trying to work out the computer format conversion in a totally digital way.)

However, most of the drawings contain more detail than can be displayed on a single DVI screen, so we will need to digitize the drawings at higher resolution, divide them up into smaller frames, and display them in a window which the user can pan or zoom to see what he or she is looking for. Some numbers for this:

- An 8 in. x 10 in. line drawing:
- Digitizing at 240 pixels/inch generates (without compression):
  4.6 megabytes at 8 bpp
  or (with DVI 16-bpp lossless compression):
  1.4 megabytes
- Digitizing at 100 pixels/inch (without compression):
  1 megabyte at 8 bpp
  or (with DVI 16-bpp lossless compression):
  0.3 megabytes

Considering that we have about 75 such drawings to handle, we can see that doing this at 240/inch resolution with 8 bpp mode and no compression would take 345 megabytes of storage, but that would reduce to about 100 megabytes if we use the 16-bpp lossless compression algorithm (which compresses about 7:1 for line drawings, but grows 2:1 because it is 16-bpp). Similarly, using a lower resolution of about 100/inch would take only 22.5 megabytes of storage using compression.

Note that the line drawings are black and white — the actual information content is only 1 bpp — if we operated in that kind of mode, the data requirement could be reduced by another factor of eight. DVI Technology must have at least 8 bpp for display, but it could handle 1 bpp image data if some custom microcode was written to expand a 1 bpp file up to 8 bpp for display. This would just be another type of decompression. For an application which needed to handle a larger number of line drawings (much larger than 75), an investment in custom microcode programming would be worthwhile; however, for this project it looks like we have enough storage space to get away with using the already-existing DVI lossless compression for our line drawings at 100 pixels/inch.

Another thing we still need to figure out is the display of the illustrations along with the text. There is a lot of activity in the industry devoted to display of text and illustrations from CD-ROM using other display systems. We should look at that field to see whether there is any display programming already on the market that we could use here. If it has already been done, there's no reason to be reinventing the wheel. In general, when a new technique is needed for an application, we should look around to see what may already be available rather than always assuming that we will do it all ourselves. As the users of DVI Technology grow in numbers, there will certainly be libraries of programming techniques available for use by others. However, at this early stage, the only source of such things is the DVI runtime software itself. The DVI software does not have any routines already tailored for the display format we will need here, so we will assume in this case that we will do it ourselves.

Indexing

Yet another aspect of the display of text and illustrations is indexing to make it easy to find subjects and to move horizontally from one class of data to another. This also is an area where there is a lot of work elsewhere in the industry using other types of display. Several companies have *retrieval engine* software for accessing large bodies of data from CD-ROM. We can expect that those techniques will be ported to DVI Technology, but it hasn't been done as of this writing. Therefore, this is another area where we will have to do some original programming to set up a system of indexing. Of course there will also be a lot of work to actually create the index tables that our software will use.

Data needs of motion video

The motion video also needs to be evaluated for its data needs. The 2 minutes of full-screen video is going to take:

$$150,000 \text{ bytes/second} \times 60 \text{ seconds} \times 2 = 18 \text{ megabytes}$$

and the 5 minutes of video in a 190x160 window (which is half-screen) will take 22.5 megabytes. Looking back at the data total, we now have 64 megabytes of existing data plus 22.5 megabytes which will come from digitizing our line drawings plus 40.5 megabytes for motion video — we're up to 127 megabytes.

Data needs for photographs

But how much data for the photographs? Most of these are photos taken of a 512x480x16- bpp DVI screen, so the thing to do is to recreate those screens with DVI Technology and save them as compressed image files. The rest are photos of live scenes, so these will have to be digitized — 512x480 should be all right for those. Then we will have 50 512x480x16-bpp image files in lossless compressed format; if we assume an average compression factor of 2:1, this will add up to:

$$512 \times 480 \times (2 \text{ bytes/pixel/2 for compression}) \times 50$$

or about 12.3 megabytes.

Now our data total is up to about 140 megabytes. We are still far from the CD-ROM limit, so we could look for more things to add to the application! However, we may want to look at our costs before considering anything more — can we afford to do what we have already outlined? We won't actually try to cost out the application here, but we can summarize the project tasks that have already been identified:

Nonprogramming Tasks
Screen design and artwork
Video production of intro and installation videos
Compress motion video
Digitize and compress line drawings
Digitize and compress still photographs
Convert VAX text files to MS-DOS
Convert Macintosh text files to MS-DOS
Create the index

Programming Tasks
Main program
Opening screen and menu
Display of text
Display of line drawing with pan & zoom
Display of photographs
Running introduction video
Running installation video with simultaneous graphics
Running DVI demonstration programs
Mechanics of indexing and display of index

The above is a fairly typical list of tasks for an interactive video application design. There are needs for artwork, and for video production. There is a lot of data preparation involved — information is seldom available in the right format. And there is programming, including special programming requirements where the particular feature we want has not been done before and will have to be developed from scratch. However we should realize that as any software-based system becomes more widely used, a bank of software techniques always grows up, so that application designers and programmers will have a growing bag of tools. Of course, the usual direction that things take is for applications to also become more sophisticated at least at the same rate as the tools improve, so the design task never really gets easier!

So far in this chapter we have only covered the things which must be done before you can start programming. Now we are ready to program, so let's look at some of the programming aspects of creating the application we have described. We will not do the whole program here, but we will look at some further techniques needed in addition to the techniques in the example program we started in Chapter 9.

Planning of System Resources

The first step in starting programming is to consider the application's needs for the system resources and set up how we will assign those resources to the different parts of the application. We have already considered the mass storage needs of the application and decided to use CD-ROM as the primary storage medium. However, we also should think about how to deploy the system's processor resources, the CD-ROM data rate, and VRAM. Let's take the processors first.

For this application there is not much problem with processing resources except when playing the installation video where we will want to be writing additional information on the screen during the video. However, since the video for this sequence is already half-screen, we can simply ask to have it compressed using only about 60% of the 82750 PA cycles for decompression, which should leave more than enough cycles to draw some graphics and text during the motion video. Otherwise, the standard arrangement of processing provided by the default DVI system should suffice.

Similarly, the only time we will be doing more than motion video and audio alone from the CD could be during the installation video where we might wish to include the graphics commands as underlay data in the AVSS file. Even in this case, the data will be so small compared to the video that there is not a problem. In fact, we might choose to not include the graphics data that way because it just adds another step in the production of the video file — for the small amount that we will use, it would be better to store it separately as a file of commands with frame numbers attached to indicate where something should be done. Then we can set up an AVSS hook routine which will read frame numbers while playing the video and execute graphics when the playing frame number matches the number stored with the graphics.

The third system resource we must consider is VRAM, which deserves a lot of discussion. Since the users of this application are DVI developers and we can expect them to have access to a DVI development system, we can rely on at least 2 megabytes of VRAM being available because 2 megs is the minimum for a development system. Let's see how this will work out. We'll make a table of the VRAM requirements for each of the different ways that this application will operate:

VRAM Requirements

(1 page = 64 kbytes)

Opening screen (512x480x16):
Screen bitmap	7.5 pages
Off-screen to load image	7.5 pages
Cursor image and save	0.1 page

Motion video without graphics (256x240x9):
Frame que (6 frames)	6.6 pages
Compressed data buffer	3.3 pages

Motion video with background video and graphics (256x240x9):
Frame queue (6 frames)	6.6 pages
Compressed data buffer	3.3 pages
Save background video	1.1 page
Cursor image and save	0.1 page

Display of scrolling text (512x480x8):
Screen bitmaps (2)	8 pages
Up and down buffers	8 pages

Display of line drawings (512x480x8):
Screen bitmaps (2)	8 pages
Compressed data buffer	1 page
Source frame buffers (4)	16 pages
Cursor image and save	0.1 page

The above table makes certain assumptions about how each of the modes will be done. In some cases there are alternative approaches which could use less VRAM for a small trade-off. For example, the opening screen's image could be loaded directly into the screen buffer with the display blanked by **GrBmDisplay(NULL)**, since this happens at the start of the application anyway. However, there will probably be occasions after the application has started that we would like to come back to the opening screen, and blanking out for several seconds while

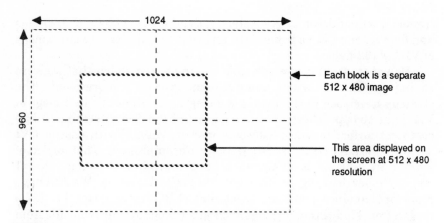

Each block is a separate 512 x 480 image

This area displayed on the screen at 512 x 480 resolution

Figure 14.1   Shows how a 1024 x 960 off-screen bitmap is used to provide a pannable display of a large image

the screen loads would not be nice. Since there is plenty of VRAM for the opening screen anyway, we might as well have the luxury of a full off-screen buffer.

Examining the table shows that the display of line drawings uses the most VRAM. Let's look at how this mode would operate and why it takes so much VRAM. To show a digitized 8 in. x 10 in. page done at approximately 100 pixels per inch, we will divide that page into four 512x480 frames, as shown in Figure 14.1. The figure shows that the four frames are slightly larger than the drawing page — this is necessary because the aspect ratio of an 8x10 horizontal page is slightly lower than the 4:3 video screen. The unused 64 pixels at the right compensate for this. (Because 512x480 resolution at 4:3 aspect ratio does not have square pixels, the 8x10 page must actually be scanned at 96 dots/inch horizontally and 120 dots/inch vertically.)

When displaying the drawing at full magnification, we will show the equivalent of only one frame of the four; however the frame being displayed actually is made up of parts from all four frames as shown in Figure 14.1. We must decompress the four frames and have them all in memory at the same time in order to rapidly create the display frame by copying. We will use the double-buffer approach for the screen so we can copy a new screen in the alternate screen buffer and then switch the display to that buffer as soon as it is ready. This way we can get smooth scrolling in each direction. This means that we need to have at least six frames worth of decompressed storage available — two for the screens and four for the source frames. In any case, it still fits into our 2 megs of VRAM (32 pages).

However, at this point someone is going to remember that we were saving the line drawings using 16-bpp lossless compression! The strategy described in the previous paragraph will not fit in 2 megs of VRAM at 16 bpp. Therefore we must revisit the decision to use 16-bpp compression for drawings — the only answer available with existing DVI software packages is to use 8-bpp Y-only uncompressed. That format of storage at 100 pixels to the inch on the original drawing takes 1 megabyte per drawing, or 75 megabytes for 75 drawings. We still have room for that since it only increases our CD-ROM storage total to 217.5 megabytes. Note, however, that loading a 1-megabyte drawing file from CD-ROM is going to take more than 6 seconds. This makes the writing of some custom microcode to handle these drawings at 1-bpp look more attractive.

The previous discussion brings up a kind of choice which will often occur when dealing with computer data from different systems. An alternative to the digitized image approach which we have been considering is to create a format conversion program that can directly convert the original vector drawing format of the line drawings into a DVI line drawing format. If there was a larger number of line drawings to be included in the application, then the software design effort for such a conversion program might be worthwhile. Furthermore, that approach could well yield more data compression than 1 bpp, and with the speed of the 82750 PA at line drawing, the display might even be faster. There are usually several alternative approaches for any format-translation problem.

Another consideration of VRAM management in a multimode setup like we need here is how we can smoothly go from one mode to another. This can be done if we arrange to have nonoverlapping bitmaps available for every change between modes. The worst case for change of mode in our example occurs between the opening screen mode and the line drawing display mode. Because it is going to take a few seconds to get a line drawing loaded before we can display anything in that mode, we probably would like to stay with the opening screen displayed until there is a drawing loaded that we can instantly switch to. However, it takes 25 pages of VRAM to set up the line drawing mode, and we are using nearly 8 pages to display the opening screen. That is 33 pages — more than our 2 megabytes, without even counting the 2 pages needed for the runtime software work space. There are several alternatives to get around this:

1. Blank out the screen with **GrBmDisplay(NULL)** and go ahead and re-use the VRAM used by the opening screen display. This will give the user a black screen for several seconds between modes.

2. As soon as the first frame of the drawing has been loaded and decompressed, switch to displaying that while we continue loading and setting up the two screen buffers for the full pan mode in the space where the opening screen bitmap was. (This will require using a similar strategy when returning to the opening screen — display a temporary image while loading the opening screen.)

3. Change the opening screen to 9-bpp mode. Then it would all fit in VRAM.

4. Use some other temporary 9-bpp screen to cover the transition — for example a screen of a soft color with the word "LOADING" displayed in the center.

As you can see, a compromise is required. We'll not decide on this one, but the point is that there will be times when you cannot get exactly what you had in mind. However, if you are clever there will usually be a list of possibilities for coming close to your desires. It's all in getting to know the limitations of your system and the different ways of avoiding those limits. (This applies to any system.)

Programming in C

Now let's look at some actual software. (Note to the non-software reader: you may want to skip ahead to Points Made by the Software Examples.) In most respects, the overall program needed here is quite similar to the overall program **main( )** used for the example in Chapter 9. However, we will modify it to add support for the mouse. Support for the Microsoft Mouse or compatible devices is built into RTX. It is set up by calling **RtxMouseInit( )**. When we initialize the mouse under RTX, a separate RTX task is activated to keep track of the mouse buttons and positions, returning values in global variables. RTX also sets up an event flag which will be set whenever the mouse is moved or any button on the mouse is pressed or released. The event flags in RTX work as described below.

Events in RTX are flags which can indicate to any RTX task that something has happened. There are certain predefined events in RTX (such as RTX_MOUSE_EVENT), and there can be other events which are programmer-defined. An RTX task looks for an event by entering the call **RtxEventWait( )**, which causes the task to stop and wait for an event to occur. The events which are valid for this call are named in the parameter list of **RtxEventWait( )**. Multiple events can be waited for by adding them to the parameter list

```
RtxEventWait(2,RTX_MOUSE_EVENT,RTX_KB_EVENT);
```

For example, the above call will wait until either the mouse or the keyboard is touched. (Note that only the task which made the call is waiting — RTX is keeping all other tasks running while one task waits for the keyboard or the mouse.)

When either keyboard or mouse is touched, the task will need to know which one it was — this is taken care of by the call **RtxTaskResumedBy( )**, which takes the name of an event as a parameter. When that particular event occurs while waiting, **RtxEventWait( )** returns, and **RtxTaskResumedBy( )** returns TRUE. Therefore, we can test the return value of **RtxTaskResumedBy( )** to find out which event it was. Neither of these functions will clear the event — if we want to clear it in order to start another wait, we must call the function **RtxEventReadClear( )**. We will use these functions in a revision to **menu( )** from Chapter 9 to add menu selection from the mouse. This is shown in Listing 14.2. However, before we look at that, we need to make some changes in the initialization to provide for the mouse. That is shown in a new listing for **initialize( )** shown in Listing 14.1.

```
/*************** Start of Listing 14.1 *************/
I16     initialize()
{
I16     x;
U16     Mask;

/* Turn off the video display */
x = GrBmDisplay(NULL);
```

```
if (x < 0) return(x);

/*  Allocate bitmap data structures  */
x = GrBmAlloc(&pBM);
if (x < 0) return(x);
x = GrBmAlloc(&pBMw);
if (x < 0) return(x);
x = GrBmAlloc(&pBMc);
if (x < 0) return(x);
/*  Describe the screen bitmap format:                    */
/*  at address 0x10000 in VRAM,512x480 resolution, 16 bpp  */
x = GrBmSetPacked(pBM, GR_BM_16, 0x10000L, SC_WIDTH, _HEIGHT,
                  SC_WIDTH);
if (x < 0) return(x);

/*  Describe the work bitmap format:                        */
/*  at address 0x90000 in VRAM,   512x480 resolution, 16 bpp */
x = GrBmSetPacked(pBMw, GR_BM_16, 0x90000L, SC_WIDTH, _HEIGHT,
                  SC_WIDTH);
if (x < 0) return(x);

/*  Describe the cursor bitmap format:                      */
/*  at address 0x1c0000 in VRAM,  512x32 resolution,  16 bpp */
x = GrBmSetPacked(pBMc, GR_BM_16, 0x1c0000L, SC_WIDTH, 32,
                  SC_WIDTH);
if (x < 0) return(x);

/*  Fill the colors[] array with 16-bpp values       */
/*  (white,yellow,cyan,green,magenta,red,blue,black  */
GrPixFromColor(pBM, GR_COLOR_RGB, 255, 255, 255, &colors[0]);
GrPixFromColor(pBM, GR_COLOR_RGB, 255, 255,   0, &colors[1]);
GrPixFromColor(pBM, GR_COLOR_RGB,   0, 255, 255, &colors[2]);
GrPixFromColor(pBM, GR_COLOR_RGB,   0, 255,   0, &colors[3]);
GrPixFromColor(pBM, GR_COLOR_RGB, 255,   0, 255, &colors[4]);
GrPixFromColor(pBM, GR_COLOR_RGB, 255,   0,   0, &colors[5]);
GrPixFromColor(pBM, GR_COLOR_RGB,   0,   0, 255, &colors[6]);
GrPixFromColor(pBM, GR_COLOR_RGB,   0,   0,   0, &colors[7]);

/*  Clear the cursor bitmap to black  */
GrBmSet(pBMc, GrBmDrawColor, colors[7]);
```

```
GrBmSet(pBMc, GrBmDrawOutline, FALSE);
GrRect(pBMc, 0, 0, 512, 32);

/* Draw the cursor in the cursor bitmap at 0,0   */
/*   cursor size is 12 x 16 pixels               */
GrBmSet(pBMc, GrBmDrawColor, colors[5]); /* red */
GrLine(pBMc,0,0,0,14);
GrLine(pBMc,0,0,9,9);
GrLine(pBMc,8,9,10,9);
GrLine(pBMc,1,2,8,9);
GrLine(pBMc,5,10,9,10);
GrLine(pBMc,2,11,8,11);
GrLine(pBMc,1,12,6,12);
GrLine(pBMc,5,13,8,13);
GrLine(pBMc,5,14,8,14);
GrLine(pBMc,5,15,9,15);
GrLine(pBMc,0,13,2,13);
GrBmSet(pBMc, GrBmDrawColor, colors[0]); /* white */
GrLine(pBMc,1,3,1,12);
GrLine(pBMc,2,4,2,11);
GrLine(pBMc,3,5,3,11);
GrLine(pBMc,4,6,4,11);
GrLine(pBMc,5,7,5,10);
GrLine(pBMc,6,8,6,10);
GrLine(pBMc,7,9,7,10);

/* Initialize the mouse   */
Mask = RTX_MOUSE_MOVE | RTX_MOUSE_LEFT_DOWN | X_MOUSE_LEFT_UP |
    RTX_MOUSE_RIGHT_DOWN | RTX_MOUSE_RIGHT_UP;
x=RtxMouseInit(Mask, 0, (SC_WIDTH-14)*MSMLT, 0,
            (SC_HEIGHT-)*MSMLT, 1, 1, mx*MSMLT, my*MSMLT);
if (x < 0) return(x);

/* Clear the screen bitmap to black   */
clearBM();

/* Display the bitmap */
x = GrBmDisplay(pBM);
if (x < 0) return(x);
```

```
return(0);
}
/************** End of Listing 14.1  ****************/
```

We will discuss only the parts of **initialize( )** which are changed for the mouse. First, we add another bitmap handle **pBMc** for a bitmap to hold a cursor image and to save the part of the screen which is currently under the cursor. This bitmap is set up near the top of 2 megs of VRAM, starting at address 0x1c0000. This takes another set of calls to **GrBmAlloc( )** and **GrBmSetPacked( )**. Then we clear the cursor bitmap to black and draw the cursor at address 0,0 in **BMc**. This is done with a series of calls to **GrLine( )**, a function which draws a 1-pixel thickness line between the two points specified in the parameters. The cursor is a small arrow pointing up—white with a red outline, as shown in Figure 14.2.

Then the mouse has to be initialized with **RtxMouseInit( )**. This function expects that a Microsoft Mouse or compatible device is already installed with its driver software. **RtxMouseInit( )** has parameters as follows:

- A flag word which tells which aspects of the mouse that RTX will respond to—a series of flags are or-ed together to tell the system to respond to movement and/or both buttons.
- The range of values the mouse will return (the constant MSMLT is defined to be 16, so the mouse returns values which are 16 times larger than the screen parameters — this factor will be divided back out later), the range is given first for the x direction and then for y (total of four parameters). Note that this range (without MSMLT) is less than the full screen to make sure no part of the cursor will go off the edges of the screen. If the cursor has to be able to go to the very edges of the screen, more elaborate routines can be written to clip the cursor when it comes near an edge.
- Two parameters which specify the ratio between mouse movement and mouse output range.
- And finally, two more parameters which specify what the mouse starting position will be.

Then **initialize( )** function finishes by displaying the screen bitmap.

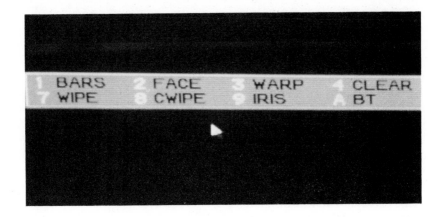

Figure 14.2   Portion of menu showing the cursor

We have repeated the entire listing of **menu( )** in Listing 14.2 below to make it easier to follow, but we will describe only the parts which have been changed to implement mouse operation. The beginning part of **menu( )** which draws the menu and writes the text in it is unchanged. However, beyond the line

                    menu_drawn = 1,

all the code is new.

```
/************* Start of Listing 14.2 ****************/
I16     menu()
{
I16     x, y, count;
GrPix   color;

/*  Load the font, if it has not been done yet  */
if (text_setup == 0) {
  x = GrFontOpen("sans.112", NULL, &font_handle);
   if (x >= 0) {
     x = GrFontLoad(font_handle);
      if (x >= 0) {
        GrBmSet(pBM, GrBmTextFont, font_handle);
```

```
        GrBmSet(pBMw, GrBmTextFont, font_handle);
            }
        }
MENU_HEIGHT);
    GrRect(pBM, MENU_X + 1, MENU_Y + 1, MENU_WIDTH - 2,
        MENU_HEIGHT - 2);

    /* Set up the text to be transparent */
    GrBmSet(pBM, GrBmTextBg, FALSE);

    /* Set up start location to draw text from array: item[] */
    x = MENU_X + 10;
    y = MENU_Y + 4;

    count = 0;
    /* Draw menu items every 78 pixels horizontally,
     *   until a null item */
    while( (count < 24) && (*item[count] != (char) ' ') ) {
        if (count == 12) {
                y += 15;
                x = MENU_X + 10;
                }

        /* Draw the item number or letter in white */
        GrBmSet(pBM, GrBmDrawColor, colors[0]);   /* white */
        GrText(pBM, x, y, item[count]);
        count++;
        /* Draw the item name in black */
        GrBmSet(pBM, GrBmDrawColor, colors[7]);   /* black */
        GrText(pBM, x + 16, y, item[count]);
        x += MENU_PITCH;
        count++;
        }
    menu_drawn = 1;
    }

/* Save under the cursor */
GrBmSet(pBMc, GrBmCopyKey, FALSE);
GrCopy(pBMc, 20, 0, 14, 20, pBM, mx, my);
/* Display the cursor with a transparent copy */
```

```
GrBmSet(pBM, GrBmCopyKey, TRUE);
GrBmSet(pBM, GrBmCopyKeyColor, colors[7]);   /*  black  */
GrCopy(pBM, mx, my+1, 12, 16, pBMc, 0, 0);
GrBmSet(pBM, GrBmCopyKey, FALSE);

/*  Now respond to the mouse   */
while (TRUE) {
      /*  Now wait until the user does something  */
     RtxEventWaitCrit(1,RTX_MOUSE_EVENT);

      /*  Put back under the cursor  */
     GrCopy(pBM,mx, my, 14, 20, pBMc, 20, 0);

      /*  save the mouse variables locally  */
     RtxGet(RtxMouseMask, &mm);
     RtxGet(RtxMouseButton, &mb);
     RtxGet(RtxMouseX, &mx);
     RtxGet(RtxMouseY, &my);
     RtxTaskEndCrit();

     mx /= MSMLT;
     my /= MSMLT;

    RtxEventReadClear(RTX_MOUSE_EVENT);

    if ((mb) && mouse_in(MENU_X, MENU_Y, MENU_WIDTH,
                     MENU_HEIGHT, mx, my)) {
        /*  Test whether the mouse is on an item  */
        if ( ( ((mx - MENU_X - 10) % MENU_PITCH) <
           ITEM_LENGTH) &&
          (((my - MENU_Y - 4) % MENU_LINE_PITCH)
          < MENU_LINE_HEIGHT) ) {
             x = (mx - MENU_X - 10) / MENU_PITCH;
             y = (my - MENU_Y - 4) / MENU_LINE_PITCH;
             if (y) x += 6;
             return(x + 1);
             }
        }
     /*  Quit if clicked in the lower right corner  */
     if (mb && (mx > 450) && (my > 400)) return(0);
```

```
    /* Save under the cursor */
    GrBmSet(pBMc, GrBmCopyKey, FALSE);
    GrCopy(pBMc,20,0,14,20,pBM,mx, my);
    /* Display the cursor with a transparent copy */
    GrBmSet(pBM, GrBmCopyKey, TRUE);
    GrBmSet(pBM, GrBmCopyKeyColor, colors[7]); /*  black */
    GrCopy(pBM, mx, my+1, 12, 16, pBMc, 0, 0);
    GrBmSet(pBM, GrBmCopyKey, FALSE);


    }
}
/************* End of Listing 14.2   ******************/
```

The first thing we must do to get a cursor on the screen is to save the area of the screen which will be covered (and overwritten) by the cursor. We do this with a **GrCopy( )** call, copying a 14x20 block of the screen at cursor location **mx,my** to the cursor bitmap **BMc** at address 20,0. The reason for offsetting to 20,0 in **BMc** is because the cursor image was placed at 0,0 in **BMc** and we do not want to disturb that. Then we can copy the cursor image to the screen. However, this must be a transparent copy so that only the cursor arrow shape appears. To do this we set attribute GrBmCopyKey to TRUE with **GrBmSet( )**, and then set GrBmCopyKeyColor to black, which is the background color for the cursor image. Then, when we do the copy, only the cursor colors different from black will copy. Note that the cursor is copied to a y-location **my + 1**; that is because we want the tip of the cursor arrow to be just below the pixel pointed to by the cursor. Now we enter a loop to respond to mouse movement. The first thing in the loop is a **RtxEventWaitCrit(1, RTX_MOUSE_EVENT)**, which will not return until the mouse either is moved or a button is clicked. (However, while this task is waiting for that, RTX is letting any other active tasks continue to run.) When the user moves or clicks the mouse, **RtxEventWaitCrit( )** returns, and the first thing we do is put back the image which was beneath the cursor, so that no matter what happens, the screen image will be intact.

Then we read the mouse global variables into local variables to freeze them at the current values while we interpret their meaning. At this point we can end the critical section by calling **RtxTaskEndCrit( )**. The mousex and mousey variables which report the mouse position are divided by MSMLT to scale them down to the screen size. These are captured in the local variables **mx** and **my**. Then we call **RtxEventReadClear( )** to make sure that this event is removed from the queue.

We set the reading of the global variables into a *critical section* of code by using the call **RtxEventWaitCrit( )** before and the call **RtxTaskEndCrit( )** after reading the variables. The purpose of a critical section is to make sure that no other RTX task can interrupt an operation before it is completed. Code inside a critical section cannot be interrupted by another task, even one of higher priority. In this case, that is necessary to make sure that all four mouse variables are read for the same condition of the mouse. If it were not a critical section, it would be possible for the mouse task to run in the middle of reading the variables, and maybe we would get a mix of readings between two different mouse positions. (Because critical sections effectively turn off all other tasks, they should be kept as short as possible.)

Now we can test the variables **mx** and **my** to see what the mouse position is. We test whether the mouse is located within the menu using the function **mouse_in( )** (explained below), and since that only matters when a button is clicked, we also test for a button with the variable **mb**. If a button is down AND the mouse is in the menu, we then test to see if it is on a menu item. If it is on an item, we calculate which item it is. This value is returned.

If any of the above conditions is not true, we proceed and test whether the mouse has been clicked while in the lower right corner. If this is true, we return the value 0, which tells **main( )** to quit. If it is not true, we proceed to get ready for another pass through the loop. We again save the screen where the cursor will go. (Note that the variables **mx** and **my** have been updated from the mouse variables, so if the mouse moved since the last time, we are saving a new location.) Then we copy the cursor image to the screen and go back to the start of the loop.

Listing 14.3 shows the function **mouse_in( )**, which tests whether the mouse position (or any x,y position) is within a specified rectangle.

```
/***************** Start of Listing 14.3 ***************/
/*  check whether mouse is in a given rectangle  */
/*  returns TRUE if it is, FALSE if not          */
int   mouse_in(x, y, xlen, ylen, xm, ym)
int x,y,xlen,ylen,xm,ym;
{
if ((xm < x) | (ym < y) | (xm >= (x+xlen)) | (ym >= (y+ylen)))
```

```
    return(FALSE);
return(TRUE);
}
/***************** End of Listing 14.3 *****************/
```

This function takes a rectangle specification and x,y coordinates **xm** and **ym** as parameters. It then does several compares to determine whether the coordinates are inside the rectangle. If they are, TRUE is returned, otherwise FALSE is returned.

Under RTX, it is not appropriate to wait for things to happen by polling with the system CPU. Polling with the CPU in one task will suspend all other tasks until the polling completes, because there is no way for the RTX scheduler to get in and switch tasks. This can be avoided by using **RtxEventWait( )** instead of CPU polling. That is a good approach for events like the keyboard or mouse, but it may prove awkward for something like waiting for the PA to finish the display list. However, we have already seen that the function **GrWaitDone( )** will do that for us.

A strategy to break up tight CPU loops is to insert a **RtxEventWait(1, RtxTimeout(0))** call into the loop. That will intro-duce a negligible delay, but it will create an opening where higher priority tasks can still get time to run. (Note that the **RtxTimeout( )** function used above is actually a delay timer which operates in multiples of 1/30 second.) Using **RtxEventWait(1, RtxTimeout(time))**, where **time** gives the number of 1/30 second intervals to wait, is a convenient way to insert small delays into a program.

Points Made by the Software Examples

We have shown how one programs a DVI application to use the mouse to make selections from a menu. The purpose of these examples is to show programmers that the features of the DVI C library make this relatively easy to implement.

The full programming of the application which we planned earlier in this chapter is beyond the scope of this book. A full application is simply built up using some of the general functions we described earlier and using many other functions in the DVI runtime software that we did not demonstrate. The DVI runtime software is a starting environment to which any serious application design group will add their own catalogue of special functions tailored to the particular class of applications that the group works with. At the same time, the DVI runtime software itself is being expanded to add further unique functions to exploit the many other features which are inherent in the DVI hardware. Over time, DVI applications will become more powerful, while DVI programming will become easier.

Summary

The design of an interactive audio/video/computer application is a complex task. It requires careful management and planning to achieve a successful result at reasonable cost. A typical application project will require artwork, audio/video production, data preparation, and programming. All of these diverse skill areas must work together to achieve the application objectives.

Programming of an application in DVI Technology is done in the C language or with an authoring system; the DVI C runtime software package provides the generic functions needed to access any of the DVI Technology capabilities. These must be combined with other programming elements such as data base functions, text processing functions, or mathematical functions to suit the needs of any particular application. As interactive audio/video applications become more common, more higher-level capabilities for application design will become available.

15

Authoring Systems

The discussion in the previous chapter focused on creating DVI Technology applications by programming in the C language. But many people who need to make applications don't have the capability to program in C, or they would like to have a development approach that is quicker and less expensive than C. This chapter is about less technically intensive, quicker, and less-expensive ways to produce applications by the use of an authoring system.

Authoring is the task of creating a multimedia application. It includes not only the generation of the application structure (programming), but also the selection of *content material* — text, images, audio, video, and any other data which supports the application and will be included with it. The field of authoring is already highly developed; particularly in support of computer-based training (CBT). Authoring hardware and software is available from many vendors and in many configurations. Industry surveys indicate that more than 70 companies are in the authoring systems business.

The capability of existing authoring software ranges from limited systems which only support text-based applications running on standard PC display formats, to very sophisticated systems for developing applications to run on systems containing a wide range of special peripherals. Such peripherals include video discs, high-resolution graphics and video boards, audio boards, audio/video capture, and touch screen displays. Software prices range from a few hundred dollars to many thousands.

General Authoring Requirements

An important aspect of authoring relates to the skills required of the person(s) doing the authoring. Most systems include some kind of specialized author interface to simplify the author's task. In the more elaborate systems, there may be more than one author interface, with each one optimized for the skill of a particular class of author. This is valuable because a large application project will require many different disciplines, and an authoring system needs to be usable by all of the team members, whether they are programmers, graphic artists, instructional designers, video producers, or project managers. The authoring system is where the work of these different people gets integrated into the application.

Some authoring systems address the issue of ease of authoring and ease of learning the system as their most important features, often heavily trading off capability and flexibility in order to simplify authoring. That is not acceptable to most people producing applications professionally — they want a capable and flexible system and they are prepared to make a reasonable investment in learning a good system. After all, authoring is their livelihood, and they will do what it takes to gain the skills needed to perform it well. The important thing is that the authoring system should *assist* the author in doing a creative job, while removing as many of the mundane housekeeping details as possible. However, it should not put limits on the author's creativity.

Many authoring systems use a mouse-driven graphical interface for selection of functions. That is a good approach when dealing with graphics or images, but it is less effective for creating and managing the structural aspects of an application, such as sequencing, branching, conditionals, or data. Some systems try to do that anyway, with mixed results.

An alternative to a graphical interface is a command-oriented interface, where the author types his/her input from a keyboard. This implies some kind of language or command set which the author will use, usually referred to as an *authoring language*. The author's typing is interpreted by the system, and commands may be either directly executed to produce the desired result, or they may be compiled into the application. A well-conceived authoring language can provide access to every aspect of the system's capability, and may be designed to have several levels of access, ranging from high-level commands which perform complex functions with one command, to detail-level commands which get right at the guts of the system.

The best approach is probably to have both graphical and language-based interfaces. Many of the users of that authoring system can do all they need with the graphical interfaces, but the language is accessible to programming-oriented users who can create the unique functions needed to make really effective and competitive applications. In fact, graphical interfaces for a language-based system can be written in the language itself — so it can have the best of both worlds. In this kind of environment, skilled users can create their own unique author interface that best suits their application objectives.

Special Needs of an All-Digital System

Most of the authoring systems on the market today were developed for systems where the audio and video came from analog peripherals interfaced to a computer, most commonly a videodisc player. This has constrained the authoring approaches used by these systems — for example, most authoring systems for videodisc use do not deal with any of the tasks of creating the video or audio material — they assume you will produce and press the videodisc before you begin any authoring. Similarly, videodisc authoring systems do not provide for any video manipulation features because the analog hardware can't do that.

An all-digital delivery system, such as DVI Technology, inherently supports a different and much better approach to authoring. You can begin authoring at the start of the application design process, and *simulate* the application while you are building it. If you intend to publish the application on a CD, there is no need to press the CD until the authoring process is all done—the application can be fully simulated and tested on the digital hardware of the development system.

The simulation process requires that the authoring hardware be able to store all of the application content on-line so that the full application can be run before any CD is pressed. In fact, the system

really has to store more material than will be used by the final application just so various alternatives can be tried out during the development
process.   This means that storage capacity in the gigabytes is not
unusual for an all-digital authoring system. It can be in the form of hard
disks or WORM drives, although large hard disk drives are probably the
preferred choice right now.

The enhanced flexibility of an all-digital system makes possible a
much wider range of application for such a system. Thus it becomes more
important that the authoring software for these systems should support
creation of many more types of metaphors which will be required by all
the different types of applications.   In other words, the authoring
software should itself be very flexible and not limiting to creativity.

DVI Authoring Software

In the following pages we will discuss two different authoring packages
that are on the market for DVI Technology.   These are the only ones
available at this writing, although there are others under development
by several companies.

One DVI authoring package is Authology®:Multimedia, which was
developed by and is being published by CEIT Systems, Inc. of San Jose,
CA. Authology uses a mouse-driven menu and window author interface
that requires minimum (not zero) programming experience from an
author.   An author for Authology should understand programming
concepts, but does not have to know any specific language.

The other package we will discuss is MEDIAscript™, which was
developed by the author of this book and is published by Network
Technology Corporation of Springfield, VA. MEDIAscript is a language-
based system which uses a text author interface. The MEDIAscript language is easy to learn, and can be used effectively by non-programmers.
However, with the powerful MEDIAscript language, a programmer can
write unique features and templates or interface modules which are
readily usable by people with no programming experience.

We will describe how each of these packages is used. The purpose of
the discussion is to help you understand how authoring software can
facilitate authoring, and how these two very different approaches
compare.

Authology: Multimedia

The author interface of Authology requires a two-monitor setup. The command monitor has a text-based window user interface which is controlled by the mouse. The authoring module for Authology is called **authed**, which is used to create, edit, or test applications or application segments. A second module **authrun** is a run-only package, which can present applications, but it cannot edit them. The run-only package has a special license so that it can be distributed economically with finished applications.

When you start **authed** the DVI screen shows a color title image and the command screen clears to a pattern background except for a menu bar across the top of the screen. The menu bar items are

```
File        Edit        Search    Window    Run
```

When you place the mouse cursor (a white block) over one of the menu bar items and click a button to make a selection, a pull-down sub-menu appears. To load a previously-created application for editing, you would click on **file** and then select **open** from the drop-down menu. At this point a dialog box will open at the center of the screen, listing files available in the current directory. You can either select from the file list or type a name in a box provided for that purpose. You can also change the current directory by clicking in another box provided for sub-directory selection. Once a file is selected, it is opened and the command screen returns to the blank background. The application can be run by going to the **Run** menu and selecting **Go**.

The menu-and-window metaphor described above is used throughout Authology. All functions are accessed this way, and there are multiple levels of menus, dialog boxes, and windows so that you can reach any particular detail aspect of your application. Very little typing is required to use Authology.

The structure for an Authology application is stored in a special binary file which is created as you are entering your application. This file becomes part of the finished application package and will be interpreted when you use **authrun** to present your finished product.

The structure file can be viewed or edited only with the **authed** module. It may be viewed by selecting **Procedure** from the **Window** drop-down menu. This opens a window which displays the instruction list for this part of the application (an application can contain many procedures). Each line in the procedure window can have a name (a label), and it then has an optional condition statement (if . . .) followed by an action instruction (then . . .). There are many types of instructions, which can be selected from a list in a dialog box. The list is shown below:

```
Remark
Assign
Show
Ask
Loop
Case
Goto
Call
Return
Execute
Exit
```

Some instructions require parameters; when these instructions are selected, you are automatically presented with a dialog box on the command monitor to select the parameters.

The **show** instruction is particulary important — it is used to present text, graphics, images, audio, video, or combinations of all. **Show** has one parameter, which is the name of a **Panel**. An Authology panel is a collection of objects which will be presented at the same time. One screen can contain one or more panels, presented sequentially.

Panels are created using the **Panel Editor**, which comes up when you select **Edit** from the dialog box which appears when you make a selection in the **Panel** window. The panel dialog box also lets you name the panel, and specify the kind of transition which will be used when the panel is displayed. Transitions such as wipes, dissolves, etc. can be specified.

The panel editor uses the DVI screen — when you enter the editor the mouse cursor will move over to the DVI screen. Again the user interface metaphor remains the same — a menu bar at the top with drop-down sub-menus. However, there is a further option where certain functions are selectable from an icon menu which you can have displayed down the left side of the screen. The panel editor menu bar is

```
Panel Object Edit Text    Line   Fill  Arrange  View
```

These items allow a panel to be assembled and adjusted until the desired effect has been obtained. We will not describe them all here, but the list of objects is the most significant for an understanding of the power of the panel editor. The items in the objects sub-menu are

```
Text
Rline
Line
Rectangle
Rrect
Circle
Ellipse
Polygon
Image
Audio
Video
Animate
Delay
Button
Field
Input
```

The function of most of these items is obvious from the names, but a few additional comments are in order. For example, when you select a **Text** object, you get a dialog box on the command screen where you should type the text content. When you click OK in the text dialog box, an adjustable window opens on the DVI screen so you can set up an area to contain the text. With various combinations of mouse and keyboard controls you can set up both the text and its location and format on the screen. You also can select font, style, and colors for the text, although these can only be set once for an entire text block. Colors are selected from a pull-down menu window which contains 256 color choices. Any of the color choices can be redefined by using a color mixing window which can be called up, so that all the colors for the current DVI screen mode are available.

The graphics objects, **Line, Rectangle, Circle, Ellipse,** and **Polygon** activate an object on-screen for which you can position, size, and select colors with the mouse. The **Image** object has a dialog box for selection of the image file to display and a box on the DVI screen so you can crop and position the image. **Audio** and **Video** are both set up from dialog boxes where you can select files and attributes. In the case of video, a box appears on the DVI screen so you can crop and position the video.

The **Animate** selection allows one or more objects to be dynamically moved on the screen, following a path that can be defined as a series of connected straight lines. Animation from an array of cells (frames) is also a choice for the **Animate** object. The **Button** object allows you to define an object (either visible or invisible) that represents the hot spot for a mouse selection.

The **Field** object is for user input from the keyboard, and the **Input** object selects either keyboard single-stroke or mouse input with optional timeout. If the user does not respond during the timeout period, a specified action will occur.

Without going into all the other items in the panel editor, you can probably already see that it is a powerful capability for building screens and controlling their presentation.

Authology supports 8-bit, 9-bit, 9-Y, or 16-bit screen modes in 256, 512, or 768 horizontal resolutions. It also has a question-answer-response and student tracking capability which comes from the CBT heritage of CEIT Systems, Inc. This latter capability also is controlled from the windowing author interface. It has capability for author-defined variables, arithmetic and logic functions, and a context-sensitive help system.

Authology is a good example of a menu-driven authoring system. It provides access to most DVI capabilities with an easy-to-learn authoring interface. It has already been used to create a number of applications in different fields, and has proven itself to many users. Although it does have a lot of flexibility, authors still come up with things which the programmers of Authology did not think of, and therefore they do not appear on the menus anywhere. For that situation, Authology offers a C-language interface so that a C programmer can write the dewired functions in a format that is linked into the Authology runtime enivronment. This feature should make possible the inclusion of many sophisticated capabilities in an Authology application.

MEDIAscript

Another example of an authoring system is MEDIAscript by Network Technology Corporation. MEDIAscript is a language-driven system with a very simple text-based author interface. That configuration contains potentially more power than Authology, but it takes more skill from an author to use it, because the author has to learn the language. However, if a language is sufficiently powerful, once an author has learned it, a language may become easier to use than a graphical interface because all parts of the application are accessible and controllable from one place — the language editor. There is no need to go through menus and windows — a few keystrokes will get you to anything.

MEDIAscript can operate with either one or two monitors on IBM PS/2 or PC AT compatible systems equipped with the ActionMedia™ boards. It supports keyboard, mouse, and touch screen user input. In the one-monitor situation, it can be set up to automatically switch the monitor between DVI-only mode when testing or running the application, to VGA overlay or VGA-only mode when editing the application code. The authoring module for MEDIAscript is called **ms**, and the run-only module for end-user delivery is called **msr**.

The MEDIAscript language contains approximately 80 commands with easy to understand names like **fill, image, text, audio, video**, etc. Table 15.1 shows the complete command set. Most commands have abbreviations, often only one letter (the list above abbreviates as **f, i, t, aud, vid**), and they also have defaults so that standard actions can be commanded by very simple syntax. At the same time, most commands have a list of optional parameters which can be added to accomplish many variations of the command.

A good example of command defaults and options is the **image** command (abbreviation **i**). If you type **i** and the name of an image file, that image will be loaded to an off-screen area of VRAM and nothing else will be done. However, you can optionally specify where to put the image off-screen with the option /**x,y** (with **x** and **y** being the pixel coordinates in a large off-screen work space). Other options allow the image to be loaded directly to the screen instead of off-screen, or to pre-load images that are in a different format from the currently-displayed screen.

Another example of command defaults and options is the **wipe** command. Using the command **wipe** by itself will wipe the most recently-loaded image to the screen using a left-to-right wipe and centering the image on the screen. However, if you add the option /- it will wipe from right to left, or if you use /**v** or /**v/-** you can choose between two vertical wipes. If you don't want the image to be centered on the screen, or if you don't want to use the entire image, you can also

Table 15.1 MEDIAscript Commands
(Abbreviations capitalized)

Multimedia Display	Developer Tools	Program Control	Graphics	File Managing
AUDio	BUF	.	ARrow	CD
BT	EDit	$*	CLS	COPY
Image	Help	BItmap	CLO	DELete
PAN	MEMory	ENDIF	CLM	DIR
WIPE	Paint	EXECute	ELlipse	ERASE
VIDeo	VARS	Goto	FEllipse	REName
	VERsion	IF	Fill	
		INNER	FPolygon	
Image Processing		INSTall	Line	**User Input**
BLEND	ROtate	LOOP	POlygon	INput
BLINK	RP	PAUSE	Rectangle	Key
BRIght	SATurate	Quit	Text	MOUSE
CLUT	SAve	RSTtimer	Vga	TOUch
CONTrast	SCALE	SCreen		
FLood	SKEW	STR		**Data Base**
GRab	TASK	Set		ASSIGN
ICOPY	TERM	WAIT		SEArch
MONO	TINT			TYpe
MOSaic	WASH			WRITE
Move	Warp			

optionally specify what part of the source image to use and where to put it on the screen.

The image-loading command is separated from the transition-to-screen command because many situations will require some action from the user before an image is displayed. There are commands which can accomplish this using either keyboard (**key**), mouse (**mouse**), or touch screen (**touch**). With the mouse, a default cursor is automatically displayed, or you can specify an image file to be used as a cursor. The **mouse** command also has a feature where you can set up a journal file containing mouse movements and button clicks which can be played back on command.

All the MEDIAscript commands which deal with images operate on rectangular blocks of pixels, which can be specified at any size or position. Image and graphics rectangle specifications take four numbers (x, y, xsize, ysize), which can be pre-specified and used repeatedly, or they can be specified separately for each command. So, for example, the graphics drawing commands **rectangle, fillrect, ellipse, fellipse** take rectangle arguments along with a color value to use for the drawing.

Colors are specified by name (white, black, blue, red, etc.), by RGB value, or in hexadecimal, and they can also be pre-specified and re-used or they can be given separately for each command. In 9-bit mode, colors may be identified as either CLUT colors or video colors.

There is a built-in editor in MEDIAscript for creating programs in the MEDIAscript language (the **edit** command). Programs are saved in *command files* which have the filename extension **.ms**. All commands except **paint** and **edit** can be used in command files. Programs can call other programs, nested up to 9 levels deep, so very elaborate structures are possible. The editor also has the feature that a command line can be executed without leaving the editor, so that lines of commands are very easily debugged while they are being created. There is also a special editor for creating simple text bullet charts, which do not require any programming at all — just type in the text contents.

For working with images, MEDIAscript has a simple mouse-driven paint tool built in (**paint** command). With this tool, you can define rectangles, move or copy rectangles, draw straight lines or rectangles, touch up pixels, mix colors, or freehand draw. The results can be saved as uncompressed or compressed image files in either 9-bit or 16-bit size. The paint tool is very convenient for creating and modifying the images or image parts which will be used in an application.

MEDIAscript provides commands for many image processing functions which work on rectangles (**bright, contrast, flood, mono, mosaic, rotate, saturate, scale, skew, tint,** and **warp**). These can be used dynamically in an application, either singly or iteratively, or they can be used for image preparation in combination with the paint tool described above.

MEDIAscript also contains a set of commands for DOS file management (**cd, copy, dir, delete, rename, type,** and **write**), which operate without leaving MEDIAscript. Text files (including command files) can be created from data produced by an application, and they can be read into an application and searched for specific information. Limited data base capabilities are possible with the language.

The VGA display is supported with line and rectangle graphics, text, and color selection. Keying between VGA and DVI on a single monitor is also supported. The **exec** command allows other programs to be run after quitting MEDIAscript, but with return to a specified point in a MEDIAscript application when the other program ends. No DOS memory overhead is needed for this function, so other programs of any size can be included in a MEDIAscript application.

Any MEDIAscript command which manipulates pixels can be set up by the **dup** command to repeat a specified number of times with incrementing or decrementing of any of the command's arguments (including colors). This type of iteration is very fast, with the speed in most cases limited only by the 82750 PA's speed.

MEDIAscript also supports author-defined variables for integer values, colors, rectangles, or strings. Arithmetic and logic is also available, and a complete set of string-manipulation functions is included.

With the MEDIAscript language, it is possible for a person skilled in the language to write template modules and tools to simplify authoring so that less-skilled people can use the system without the need to learn the language. This approach allows application developers to build their own authoring interfaces for their special needs — without any loss of flexibility. Network Technology Corp. is working on tool sets and template packages which will offer many of the basic capabilities. An application developer can start with these tools and templates and customize them where necessary.

MEDIAscript is a very powerful application development environment which anyone who makes even a small investment in learning can utilize to access the full capabilities of a DVI Technology system.

Summary

Authoring systems are needed to open up the capabilities of a complex technology to lesser-skilled users and to make the job easier for skilled users. This is an important step in reaching broader application of a new system. We have reviewed the general principles of multimedia authoring and shown two very different examples of systems that have been developed for DVI Technology. Both of these systems are in use today and are helping to build the market acceptance of DVI Technology by making applications easier to develop. As the technology and the markets continue to grow, the two systems described here will be enhanced with more features, and other authoring systems will be introduced to further speed the growth.

16

Thinking about the Future

Digital video on a personal computer is a technology that has just come out of the starting gate. While the fundamentals that support this technology have been around for more than 10 years, it is only now that products are coming out at price points which will lead to broad proliferation of the technology. In this final chapter I will discuss some of the indications which argue that digital video is ready for wide use on personal computers. I don't want to call this discussion a forecast — I'm not a forecaster — but I will try to identify the areas where the industry and the markets are moving.

Industry Trends

Several industry trends say clearly that digital video technology is not going to stand still — it will exploit the continuing developments in VLSI, digital storage, and the PC itself. Each of these three fields has seen many advances over the past 10 years and there is nothing to make us think the trends are anyway near ended. In fact, the pace of development more likely will accelerate over the next few years.

VSLI trends

VLSI will continue to move to still higher chip densities. The forces, which over the last two decades have led to a doubling of the number of devices on a chip every two years, are still at work. This is known as *Moore's Law*, named for Gordon Moore, the chairman of Intel, who first articulated it in 1965. Industry forecasts say that we can expect Moore's Law to be true for at least another decade, which means that in the year 2000 a single integrated circuit could contain as many as 100,000,000 devices. Today's latest production devices are based on chip feature sizes in the range of 1 micron and above and they contain up to 1,000,000 devices, yet the industry is rapidly building processing facilities to produce chips with submicron features. This will make custom chips practical with greater complexity, and they will run faster and cost less. At the same time, custom chip design technology is not standing still either. The silicon compiler technology used for designing the DVI VDP chip set, for example, is growing in effectiveness and efficiency to allow more complex chips to be designed more easily and to create devices which will make better use of the silicon area.

Therefore, the functionality for which we need two chips today will soon be able to be put on one chip. However, that is unlikely to be the way that the industry will go — no one will be satisfied to stay with the same functionality — future custom devices will use their improved design efficiency to add functionality while bringing costs down simultaneously.

Digital storage trends

Over the years, magnetic digital storage devices have increased in capacity while dropping in cost. That is continuing for magnetic devices, but magnetic storage now has a competitor in optical technology. While the leading optical storage device today — the CD-ROM — is read-only, writeable optical storage is becoming available at lower and lower price points. One can buy a read-write optical drive today, but the price is high and there are too many different standards. But this will change — today's writeable optical drives prove that there are basic technologies which can do the job, and now the industry has to find the best way(s) to package those technologies, standardize the products, and make them available from a number of manufacturers. It will happen, and the digital storage industry will take a leap from megabytes per chunk of medium to gigabytes per medium. As we have seen, digital video's appetite for storage capacity will be hard to satisfy even with gigabyte devices!

PC trends

And the personal computer field is not standing still either. We see new faster, lower-cost machines with new features being introduced almost every day. This trend also can, and will, continue for a long time. However, when talking about new, and presumably different, computer introductions, the subject of standards naturally comes up.

Standards

Engineers often look at standards as roadblocks to progress; however, I think when they are correctly conceived they become *gateways* to progress. The PC market is where it is today because of the de facto standard set by IBM in 1981 with the first IBM PC. At that time the technical community viewed it as a real roadblock because it did not embody even then what was considered to be state of the art. However, those same people also would never have forecast the momentum which grew from that standard. The best way to understand what a standard does is to look at it from a software developers point of view. New machines take new software — and a software developer has limited resources to constantly be revising his or her software to run on new machines. Therefore, a software developer is going to look for platforms for his or her products which will have the greatest market acceptance. That's not the newest machine, because market acceptance takes time.

The other side of the coin is that a new PC product is only as good as its software. Companies introducing new machines today must also figure out how they are going to get all the software it will take for their new machine to be accepted and sell. One way to do that is to design the new machine to use an existing body of software — then the software is already out there. But the software that is out there cannot exploit whatever new and unique features are in this new machine. Therefore, a manufacturer still has to get at least some software which can show off the new machine's unique aspects, to convince customers to buy this machine instead of one of the others which run the same standard software.

The point is that once something is recognized as a standard in the industry, tremendous support will grow for it because the software community will be comfortable developing for it — they know there will be a growing market for their software. At the same time other manufacturers will want to make similar products and also benefit from the software growth. This is what happened with the IBM PC — even though it was technically not state of the art at its time.

"Open" standards

A tricky aspect of standards is the extent to which a standard is *open* — meaning that the details behind the standard are revealed and can be used by others. A manufacturer coming out with a new system has a difficult choice to make between keeping the details of the system secret and therefore having a competitive advantage over others who may try to make a similar system versus opening up all the details and allowing others to make the same system in order to cause the total market to grow faster.

My personal view is to open up as much as possible and get as many other manufacturers as you can on board the standard. If your proposed system standard is basically sound, the customers and the software community will appreciate the open approach and more easily accept your system. Furthermore, because all your competitors are developing on top of your standard, a part of their development efforts will be helping you as well. (I have to admit that there are successful systems that do not use the open approach, but I'm free to state my opinion anyway.)

Vertical Markets

There are market segments for personal computer products which do not fit the model described above for standards. These are usually referred to as *vertical* markets — markets which have needs that are so special that they are not filled by standard hardware or software products. (*Horizontal* markets are more general-purpose, and would include many different customer categories.) Examples of vertical markets for software are business categories which have well-defined special needs such as (special requirements are in parentheses):

> Medical facilities (patient medical histories)
> Rental businesses (rental records)
> Dairy farm (cow records)
> Manufacturer (mfg data base)
> and many others.

Vertical markets which have special hardware needs are (basis for special need is in parentheses):

Military (ruggedness)
Consumer information terminals (need for good video)
Aircraft (reliability, environment)
and many others.

Because of the special needs of vertical markets, they benefit only from the elements of standard general-purpose products that they can use without change. These markets will more easily accept a new technology that addresses their needs because they will probably have special hardware and software anyway. They only have to consider their own existing customer base, and their own problems of transition to the new technology. From another point of view, the vertical markets are fertile developing ground for new technologies. However, any technology whose growth has been tied to vertical markets must understand the size limitations of these markets.

Video is already in use in certain vertical markets where video is one of the special needs, but today it is mostly analog video, usually based on laser video disc. However, these markets understand the benefits of digital video and many have made the decision to go into it. Typically the standards issue did not loom large to these companies — more important considerations were the cost, performance, and availability of the technology. However, to a developer and manufacturer of digital video hardware and software, the trick to vertical markets is to get them all to accept the same technology. If the same technology can satisfy a wide range of vertical markets, then the momentum for a standard has been started. This is part of the strategy for DVI Technology.

The Consumer Market

As digital video moves into larger and more horizontal markets (it will), the subject of standards becomes more important. The largest market of all is the consumer (home) market, and the consumer already knows what video is — it's television. Whether the consumer is ready for an interactive digital video product is not so clear. What is clear is that such a product would have to be priced under $1000, it would have to address applications such as entertainment and education, and it would have to be extremely easy to use. Furthermore, the consumer market requires a standard. The consumer cannot be expected to have the technical knowledge to choose between a number of competing system approaches — he or she would like all products in the same category to meet a single well-defined standard that assures proper technical performance, and consumer purchasing decisions will be based on more subjective factors. All of this will happen, but it will take a lot of consumer education and it will take some years of market development.

The matter of reaching a price point below $1000 for a digital video product bears more discussion. Today a digital video system costs in the range of $5,000 to $10,000. For that price you are getting a general-purpose computer, a hard disk, and a computer-style user interface. Those features do not need to be in a consumer interactive video product.

The consumer market is so large (if it accepts the product) that a manufacturer can afford to go to the ultimate in creative system design, custom chip design, and special hardware tooling to achieve the lowest manufacturing cost. A good example of what those things can accomplish is seen in the consumer VCR. The consumer VCR utilizes essentially the same technology that a broadcast-level VCR does, yet the consumer product costs in the hundreds of dollars and the broadcast product costs tens of thousands. But the market size for broadcast products is in the thousands of units, where the consumer market size is many millions of units. Therefore a much larger investment for design and manufacturing tooling makes sense to the consumer manufacturer. The result is excellent value for the consumer's dollars. This same situation will apply to digital video products when the consumer market comes into sight, and the under-$1,000 price point will happen.

High-definition Television (HDTV)

The world's television standards (NTSC, PAL, SECAM) are more than 30 years old. In view of the massive advances in technology in the last 30 years, it is a remarkable testimonial to the vision of the group who developed the standards in the 1950s, that their work remains viable today. However, today's technology would allow a vastly improved television system to be built. Why hasn't that happened, and what is being done about it?

The fact is that a number of improved systems have been developed in several countries, but as of this writing none have reached mass marketing. No one can take the risk of investing in mass marketing of a new TV system without the assurance that their offering will be backed by a viable standard. As we explained in the last section, we cannot expect the consumer to take the risk of purchasing a system without a standard behind it.

The CCIR has been working on international standards for HDTV, particularly for program production, rather than broadcasting. One reason for that is that an HDTV system can solve the program producers' problem of producing in a medium which is capable of translation to any of the world's television standards without degradation. You can do that if you produce in HDTV. In a sense, program production is a vertical market much the same as the vertical markets we discussed for digital video. It can afford a more expensive solution than the consumer can, and most important, program production does not require over-the-air transmission.

Another reason that broadcasted HDTV systems have not emerged is that the wider bandwidth of HDTV has made the spectrum allocation for a new service very difficult. It would be much easier if HDTV could fit into existing TV channels. This has caused many organizations to do more work to try to solve the spectrum utilization problem, and they are having some success.

And a third reason for not standardizing broadcasting for HDTV at this time is that there is a major controversy over the question of compatibility with existing television systems. If HDTV is not compatible with existing TV (meaning that existing TV receivers could somehow receive the HDTV programming), then viewers cannot enjoy the programming developed for HDTV unless they replace their present receivers with HDTV sets. This is of course an incentive for the consumer to buy HDTV (assuming the programming on HDTV is as attractive as the technical capability is), but it means that broadcasters who are now on the air with the old system will start losing their market as HDTV receivers are bought by the public. You can imagine the arguments that this leads to.

The delay in standardizing HDTV has also made an opportunity for development of ATV (Advanced TV) systems, which are compatible approaches that provide many of the advantages of HDTV but remain with the present TV channel bandwidths, and in some cases are simply enhancements to existing systems. These systems would be much less expensive to implement, and the receivers might also be less expensive, but of course the long range performance potential of the system is probably less than a full HDTV system. Therefore, there is an argument whether implementation of an ATV system would be worth it because it does not provide as much ultimate improvement in performance. These arguments go on, but they will eventually be resolved and standards will get set.

But what does HDTV mean to digital video and audio as we are considering those here? Well, the higher analog resolutions of HDTV translate into higher pixel resolutions if future digital video is going to match the performace of future television systems. Whereas we have said that 512 x 480 pixel resolution matches a good present-day TV system, we probably have to go to something like 1280 x 1024 to match an HDTV system. As you will see below, this is also in the cards for the future of PC-based digital video systems.

Digital Multimedia Advances

Digital multimedia itself is not going to stand still, either. It will soon become a standard part of new personal computers, with power applications which take advantage of digital video and audio to create a new level of user friendliness.

As market deployment of digital multimedia products begins, the development of new algorithms and new hardware will accelerate. The greatest challenge to a digital multimedia manufacturer or software provider will be to stay up with the advances in the basic technology. The best way to do this is to keep as much of the system characteristics defined in software and resist the pressure to cast things into the hardware. While it may look more difficult to do everything in software today, processors are going to become faster (sooner than we expect), and software will become more clever — meaning that tomorrow's software-based system will have more performance than today's hardware-based systems. Doing everything in software is the only way to keep up with algorithm developments.

The video compression field is a specific example of an area where we can expect massive advances over the next few years. Researchers in that field are now getting hardware which will increase by orders of magnitude their capability to experiment and get quick results. This is going to spur even further research. At the same time, results of earlier research are now going into the market, and the feedback from that experience will give new directions to some of the research.

Hardware also will improve. I have already mentioned faster processors — everything will get faster. By the year 2000, single-chip processors could be more than 100 times more capable than they are today. Any system which expects to survive in the market long enough to make a profit for its developers must have an effective hardware-growth strategy. In a software-based system this can be handled by proper design of driver software so that upgraded hardware with new drivers will still work with present application software. The driver-to-application interface design of the system must be thought out so that it can survive when the hardware below the driver interface is replaced with more powerful stuff.

At this point it is probably not a surprise to you to hear that the previous discussion embodies much of the strategizing which went into DVI Technology. DVI Technology is following the path of market development that I have outlined above. At this writing, it is still early in the game, but the gains already made, the customer acceptance already received, and the announced plans for enhancement indicate a great future.

Conclusion

The purpose of this book has been to teach the emerging technologies of digital video and audio for personal computers. As I have explained in this chapter, the potential in these technologies is enormous. It is clear that digital video and audio will have an important part in the future of personal computers and, as one industry observer remarked upon witnessing for the first time DVI Technology's realistic motion video and audio coming from a PC AT, "The personal computer will never be the same."

Appendix A

Worldwide Television Standards

This appendix presents a summary of the major parameters of the principal television systems in the world — NTSC, PAL, and SECAM. For more information about these standards, consult *Television Engineering Handbook*, Chapter 21 (see Appendix D, Reference 9). That reference will also lead you to the formal standards publications for many of the countries.

There are minor differences between the same system implemented in different countries. The information given here is for NTSC-M (United States), PAL-B (West Germany), and SECAM-L (France).

Scanning Parameters

Scanning	NTSC	PAL	SECAM
Lines/frame	525	625	625
Frames/second	30	25	25
Interlace ratio	2:1	2:1	2:1
Aspect ratio	4:3	4:3	4:3
Color subcar.(Hz)	3,579,545	4,433,619	Note 1
sc/h ratio	455/2	1135/4	Note 1

Sync Waveforms

Refer to Figure A.1 for nomenclature. All values given are nominal. Horizontal sync values are in microseconds.

Horizontal Timing (µs)	NTSC	PAL	SECAM
Line period (H)	63.55	64.0	64.0
Blanking width	10.9	12.0	12.0
Sync width	4.7	4.7	4.7
Front porch	1.2	1.2	1.2
Burst start	5.1	5.6	Note 2
Burst width	2.67	2.25	Note 2
Equalizing width	2.3	2.35	2.35
Vert. sync width	27.1	27.3	27.3

Vertical Sync	NTSC	PAL	SECAM
Blanking width	20H	25H	25H
Num. equalizing	6	5	5
Num. vert. sync	6	5	5
Burst suppression	9H	Note 3	Note 4

Color Matrix Equations	Chrominance Bandwidth (-3 dB)

NTSC

```
Y = 0.30 R + 0.59 G + 0.11 B
I = 0.60 R - 0.28 G - 0.32 B        1.3  MHz
Q = 0.21 R - 0.52 G + 0.31 B        0.45 MHz
```

PAL

```
Y =  0.30 R + 0.59 G + 0.11 B
U =  0.62 R - 0.52 G - 0.10 B        1.3  MHZ
V = -0.15 R - 0.29 G + 0.44 B        1.3  MHz
```

SECAM

```
Y  =  0.30 R + 0.59 G + 0.11 B
D_R = -1.33 R + 1.11 G + 0.22 B       1.3  MHz
D_B = -0.45 R - 0.88 G + 1.33 B       1.3  MHz
```

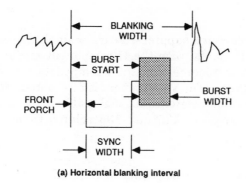

(a) Horizontal blanking interval

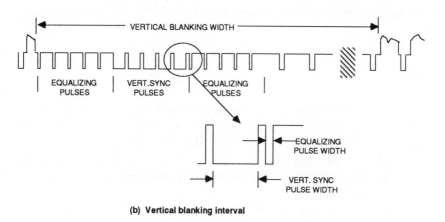

(b) Vertical blanking interval

Figure A.1   Sync waveform nomenclature

Color Modulation Parameters

NTSC
The chrominance subcarrier is suppressed-carrier amplitude modulated by the I and Q components, with the I component modulating the subcarrier at an angle of 0 degrees, and the Q component at a subcarrier phase angle of 90 degrees. The reference burst is at an angle of 57 degrees with respect to the I carrier.

PAL

The chrominance subcarrier is suppressed-carrier amplitude modulated by the U and V components, with the U component modulating the subcarrier at an angle of 0 degrees, and the V component at a subcarrier phase angle of 90 degrees. The V component is alternated 180 degrees on a line-by-line basis. The reference burst also alternates on a line-by-line basis between an angle of +135° and -135° relative to the U carrier.

SECAM

The chrominance subcarrier is frequency modulated by the D_R and D_B signals on alternate lines. At the same time, the subcarrier frequency changes on alternate lines between SC_R and SC_B (see Note 1). The color burst also alternates between the two frequencies, as explained in Note 2.

Notes

Note 1:   SECAM uses two FM-modulated color subcarriers transmitted on alternate horizontal lines. SC_R is 4.406250 MHz $(282*f_H)$, and SC_B is 4.250000 MHz $(272f_H)$.

Note 2:   SECAM places a burst of SC_R or SC_B on alternate horizontal back porches, according to the subcarrier being used on the following line.

Note 3:   Because the PAL burst alternates by 90 degrees from one line to the next, the 8.5-line burst suppression during vertical sync must be shifted according to a four-field sequence (called *meandering*) in order to ensure that each field begins with the burst at the same phase.

Note 4:   SECAM color burst is suppressed during the entire vertical blanking interval. However, a different burst (called the *bottle* signal) is inserted on the 9 lines after vertical sync.

Appendix B

Reading of C Programs

The purpose of this section is to help a reader who has some programming knowledge but who does not know the C language to understand the software examples in this book. Knowledge of another language such as BASIC or Pascal is assumed. This is not a complete exposition of C —it covers only the subset of C features that are used in the program examples in this book and deals only with reading of code. For a more complete tutorial on C, consult one or more of the references about C listed in Appendix D.

C is a *compiled* language, which means that a C program is not directly executable by the computer, but rather it must first be processed by a *compiler* program to produce an *object* file (.obj) and then further processed by a *linker* program in order to create an .exe executable program. To someone familiar with an interpreted language like BASIC, the compile-link procedure would seem to be extremely cumbersome. However, it has many advantages in terms of programming efficiency, program speed, and ability to fully exploit the resources of a computer.

A C program is created as a *source* program, usually given a filename extension **.c**. Source programs are ASCII text files and they may be created with any text editor. Any number of source files may be combined into one executable program through the compile-link process.

C is case-sensitive for variable names, statements, and function names. Normal practice is to use lower case for most names, reserving capitals for symbolic constants, which will be explained later.

Comments

Comments in C are contained inside of slash-star characters as follows:

```
/*   this is a comment   */
```

A comment may be located anywhere in a program where a space would be acceptable.

Functions

C programs are made up of functions, which are much like subroutines in BASIC or procedures in Pascal. A function definition has the following structure:

```
type    function_name ( )
{
/*  body of function  */
}
```

The function name is stated first, followed by a pair of parentheses. A function must also be given a type, which will be the type for a single return value from the function. If no type is specified for the function, C will default to type **int** (integer). The active code of the function is contained within a pair of curly braces { }. (Curly braces are used to group C program statements in many other contexts, too.) All functions are defined separately — there is no hierarchy of functions. The program flow is determined by how functions are *called*, not by how they are defined. Any function may be called from any place in a C program, even from inside of a mathematical expression. This is because a function is equal to its return value.

Declarations

The C compiler must be told ahead of time about variables, structures, and functions. This is accomplished with *declaration* statements, which set up the compiler to recognize things to be used in the source code which follows.

Overall Program Organization

The organization of a C source program is as follows:

```
#includes
global variable declarations
main ( )
other functions
```

Items which begin with the prefix character # are called *preprocessor directives* —they are commands to the compiler to do something which will assist the compilation process but will not necessarily create any executable code itself. The directive **#include** is followed by the filename of another C source program which should be read by the compiler at this time. It is usually used for so-called header files, which are lists of declarations that are needed with this source program.

In any C program, there must be a function called **main( )** —this is the top-level function and will be executed first whenever the program is run. Other functions in the program are called from **main( )** or from other functions.

Symbolic Constants

A preprocessor directive **#define** is used to define constants which the compiler will then recognize any time they are used in the source code. By convention, constant names are always in capital letters. It is good practice to define all "magic" numbers or other values at the start of the program or in header files. If the names are carefully chosen, source code becomes much more readable. Also, if a value needs to be changed later, the change only has to be made once, in the **#define** statement. For example, to define the constants TRUE and FALSE:

```
#define  FALSE   0
#define  TRUE    1
```

Variables

In C, variables must be declared before they are used, and a strict system of defining types for variables is imposed. In BASIC there is a fixed list of variable types (integer, string, single precision, double precision), but in C there can be any number of different types, and the programmer can define his or her own types. This includes setting up a complex structure of data and defining it as a type.

There is a hierarchy of variables which allows variables to be private to an individual function or global and accessible by all functions. If a variable is defined outside of any function, then it will be global — accessible by all functions in the source file. If a variable is defined within a function, then it is private and is available only to that function. Furthermore, when the function ends, the RAM spaced used by the private variables will be given up for reuse by other functions. (This is a simplified view — there are many other ways to set up variables in C, but we will not be using them in the examples in this book.)

A variable must first be declared before it is used. A variable declaration has the following form:

```
type        variable_name;
```

The type must always be stated, and the declaration statement must end with a semicolon. (All C statements must end with a semicolon.) Optionally, multiple variables of the same type may be declared in the same statement by separating the names with commas. A semicolon must still appear at the end of the statement. Variables also can be initialized by equating them to the initial value in the declaration statement as in the following example:

```
int x = 0, y = 0, i;
```

Also note that a new line character can appear in a C statement at any place where a space could be placed. The readability of C code is greatly enhanced by the proper use of spaces and new lines. For example, the statement above could be written as follows in order to introduce comments which explain the purpose for each variable that is being declared:

```
int
x = 0,      /* horiz. pixel position */
```

```
y = 0,      /* vert. pixel position */
i;          /* counting variable    */
```

Statements

All active code in C is in the form of statements. The following are
statements:

```
f1 ( );
x = start + length;
```

The first statement calls the function **f1( )**. When that function
returns, execution will continue with the following C code, in this case
that is the next line. The next line is an arithmetic statement which sets
the variable **x** to the value obtained by adding the values of the variables
start and **length**.

Operators

C provides a large number of operators for doing arithmetic, logical, data
shift, relational, and assignment functions. The subset we use in this
book is described below:

Arithmetic
```
+  -  *  / (which have their usual meanings)
%    (remainder after division)
```
Logical
```
||  (logical OR)
&&  (logical AND)
!   (logical NOT)
|   (bitwise OR)
&   (bitwise AND)
^   (bitwise EXCLUSIVE-OR)
```
Data Shift
```
>>n (shift right n times)
<<m (shift left m times)
```
Relational
```
>  <  <=  >=   (usual meanings)
```

```
==  (equal to)
!=  (not equal to)
```

Assignment

```
=   (equals)
+=  (add right operand to left operand)
-=  (same for subtraction)
&=  (AND assignment)
^=  (EXCLUSIVE-OR assignment)
++  (increment by 1)
--  (decrement by 1)
```

Functions with Arguments

A C function may receive arguments by adding a list of argument names within the parenthesis of the function name. The following is a function definition which receives two arguments of the int type:

```
void    funct ( x, y)
int     x, y;
{
/*  body of function    */
}
```

The function itself is of type **void**, which means that it has no return value. It expects to receive two **int** arguments, which are defined as **x** and **y** internally for the function. The type for the arguments is declared outside of the function braces. (Additional variables for use within the function can be declared inside of the braces.)

Pointers

For simple variable types such as **int** and **long**, the variable name directly refers to *the contents of* the variable. However, this approach will not work for more complex variable forms such as arrays and structures. In those cases, C defines the variable name to be a *pointer* to the structure or array. The pointer is simply the *address* of the start of the variable's space in memory. Two special operators are used for working with pointers:

```
*    (indirection operator)
&    (address-of operator)
```

Arrays

Arrays may be set up in C for any variable type currently defined in the source code. An array of 20 elements of type **int** would be declared as follows:

```
int       array1[20];
```

The contents of an array can be accessed in two ways. First, an array index may be used; for example, to read the **i** th element of the array above:

```
int       x, i;
x = array1[i];
```

Or it may be accessed through a pointer:

```
int       *pa, x, i = 0;
pa = &array1[i];
x = *pa;
```

In this case, the variable **pa** is declared to be a pointer to an integer. Then **pa** is initialized to the address of **array1[i]** by means of the address-of operator **&**. The contents of the array item is then read by using the indirection operator *****. Note that C knows about the type of an item being accessed by a pointer. If the pointer **pa** is incremented **(pa++)**, it will automatically shift by the correct amount to access the next integer item in memory. (There is a lot more to pointers than this, but this is all that we need for our sample programs.)

An array may be initialized (outside of a function), by including a list contained in braces in the array declaration:

```
*char    array2[ ] =
{
"ONE", "TWO", "THREE", "FOUR", "FIVE",
} ;
```

This example sets up a character (string) array **array2** which is initialized to 5 items having the values contained in double quotes in the list.

Data Structures

C provides for setting up data structures of almost any complexity by use of the word **struct**. There are no structure definitions in our example code, and there are no structure definitions in the DVI header files. The reason for this is that the DVI system internal data structures are deliberately not documented for the programmer to directly access them. Instead, data items are accessed via the **Get( )** and **Set ( )** functions, such as **GrBmGet( )** and **GrBmSet( )** used to read or write to bitmap data structures. In this way, the designers of the DVI system software are free to make future changes or enhancements in the internal data structures, but they will keep the Get/Set interface the same so that old C source code will still work without needing any changes.

Conditional Statement

The **if( )** statement provides for conditional execution of code as follows:

```
if ( /* expression */)
   {  /* execute if TRUE  */  }

/*   rest is optional   */
else {  /* execute if FALSE  */  }
```

The expression contained within the parentheses following if is evaluated to determine how execution will proceed. If the expression evaluates to nonzero (considered TRUE), then the code immediately following the **if ( )** statement will be executed. (This can be just a single statement ending with a semicolon, or it may be a group of statements enclosed in curly braces.)

When the expression in parentheses evaluates to zero, (considered FALSE), execution will jump to the code *after* the statement or group that is immediately after **if( )**. Optionally, an **else** statement can be used to insert code which will be executed *only* when the **if( )** expression is FALSE. When the **if** and **else** code is completed, compiling continues with the code that follows.

Loop Constructs

Two of C's looping constructs are used by our example code. One is the **for ( )** statement:

```
for ( x = 0; x < 20; x++ )
    { /* loop action code   */ }
```

There are three separate parts to the setup of a **for( )** loop, contained within parentheses and separated by semicolons. The first part contains an initialization statement, which will be executed before the loop begins. In the example, we set the variable **x** to zero. The second part of **for( )** is the condition which must be TRUE for execution of the loop contents to take place. In the example, this is **x < 20**, which means that the loop action code will be repeated until **x** exceeds 19. The third part of the **for( )** statement is a statement which executes after each execution of the loop action code, but before the condition statement is tested. In this case, the value of **x** is incremented. Therefore, this loop will execute 20 times, until the value of **x** reaches 20. Note that the **for()** loop makes its conditional test *before* executing the loop action code.

The other looping construct that we use in the code examples of this book is the **while( )** statement:

```
x = 0;
while (x < 20)
{
/*  loop action code  */
x++;
}
```

In this example, we have set up the **while( )** statement to perform the same operation as the previous example using the **for( )** statement. You can see that it is more complicated to use **while( )** for this simple example; however, in more complex cases, that is not always true.

While( ) also tests its condition statement before execution of the action code.

Either type of loop can be exited from within (before the loop condition itself becomes FALSE) by inserting the **break** statement. **Break** will exit the current loop and continue with the code which immediately follows the loop. Another type of exit can be obtained by using the **goto** statement. In this case, execution can be caused to jump to another line (but only within the same function) which has a *label* name which is the same as that used in the goto statement. (A label is indicated by a colon following the name.)

A further form of loop exit is the **return( )** statement, which causes exit from the current function. The parentheses on **return( )** may contain a value which to be returned by the function being exited, or if the function is void, then **return** is used without parentheses.

The Switch( ) Statement

The **switch( )** statement is used to provide a multiplicity of execution alternatives depending on the value of a single variable. A typical use is to receive a single character selection from a menu routine and direct execution to the proper function. For example:

```
int     sel;

sel = getch( );

switch (sel)
{
case `a':
case `A':
    funct_A( );
    break;
case `b':
case `B':
    funct_B( );
    break;
case `c':
case `C':
    funt_C( );
    break;
```

```
default:
    break;
}
```

The integer variable **sel** receives a character from the keyboard by calling the standard C function **getch( )**. The **switch( )** statement tests the value of **sel** and then directs execution to the proper function depending on the letter typed on the keyboard. The double **case** statements take care of either upper case or lower case keystrokes. The **break** statements are required to make sure that when the selected function returns, execution will jump over the remaining cases of the switch statement. The **default:** case is executed if the character typed does not match any of the other cases.

Standard C Functions

The C language has a large number of standard functions, some of which are used in our examples. These are explained below.

printf( )   In a standard C program, this is a function for formatted printing to the system standard output device (usually the monochrome screen). However, in DVI programs we get at the functionality of **printf( )** by using **RtxPrintf( )**, which is used in the same way. It has an elaborate syntax, but we will use only a simple example here:

```
int     x = 1234;
RtxPrintf("The value of x is %d\n", x);
```

which will print

```
The value of x is 1234
```

on the standard display. **printf( )**'s arguments consist of a format string followed by a list of values. Locations of values are identified in the format string by commands beginning with the character %. In this example, the command **%d** indicates that the value of an integer variable is to be printed where %d appears in the format string. The format string command **\n** means that a newline character is to be printed.

NULL   NULL is a C constant for an empty pointer (pointing to nothing).

fclose( )   This is one of the C functions for disk access. In the sample code of this book, most disk access occurs through the DVI runtime software functions; however **fclose( )** is used at one point to close a file that was opened by one of the DVI functions.

kbhit( )   This is a function which returns TRUE if a keyboard key has been struck since the last time the keyboard buffer was cleared. The keystroke value is not read by **kbhit( )**; another function such as **getch( )** must be used to read the key value.

Appendix C

Glossary

(Note that italicized words refer to definitions elsewhere in the glossary.)

3:2 pulldown —the technique for displaying 24 frames/second motion picture film on a 30 frames/second (interlaced) television system. One film frame is shown for 3 television fields and the next film frame is shown for 2 television fields.

access time — in mass storage devices, the time from issuance of a command to read or write a specific location until reading or writing actually begins at that location.

active pixel region — on a computer display, the area of the screen used for actual display of pixel information. (There may be additional screen area around the edges but not used to display pixel data, called the *border* region.)

additive color system — a color reproduction system where an image is reproduced by mixing appropriate amounts of red, green, and blue lights.

algorithm — A group of processing steps that perform a particular operation, such as drawing a line, or compressing a digital image.

aliasing — in *sampling*, the impairment produced when the input signal contains frequency components higher than half of the sampling rate. Typically produces jagged steps on diagonal edges. See *anti-aliasing*.

analog to digital conversion (A/D) — the process of converting an analog signal into a digital bit stream. Includes the steps of *sampling* and *quantizing*.

analog video — video in which all the information representing images is in a continuous-scale electrical signal for both amplitude and time.

anti-aliasing — the process of reducing the visibility of *aliasing* by using gray scale pixel values to smooth the appearance of jagged edges.

artifact — in video systems, something unnatural or unintended observed in the reproduction of an image by the system.

aspect ratio — the ratio of the width to the height of an electronic image. For broadcast television, the standard aspect ratio is 4:3.

Audio-Video Support System (AVSS) — in DVI runtime software, the software package which plays motion video and audio.

AVSS — see *Audio-Video Support System*.

authoring — the process of creating a multimedia application program.

bandwidth — refers to the frequency range transmitted by an analog system. In video systems, specifying the highest frequency value is sufficient, since all video systems must transmit frequencies down to 30 Hz or lower.

bank switching — the hardware technique to access a large RAM or VRAM array by dividing the array up into sections and accessing one section at a time.

binary — a digital signal having two levels, usually referred to as 0 and 1.

bit assignment — in video *compression*, the process of creating the compressed data *bit stream* from the raw output of the compression *algorithm*.

bit plane — in digital video display hardware which has more than one video memory array contributing to the displayed image in real time, each memory array is called an *image plane*; however, if the arrays have only one bit-per-pixel, they may be called bit planes.

bit stream — a serial sequence of bits.

bitmap — a region of memory or storage that contains the pixels representing an image arranged in the sequence in which they are normally scanned to display the image.

bitmap descriptor — in the DVI runtime software, a data structure that contains the parameters of a bitmap including its location in VRAM memory, its dimensions, and its pixel format.

bitmap font — a special format of a text font that contains pixel values for each text character.

bits per pixel — the number of bits used to represent the color value of each pixel in a digitized image.

blanking level — in a video signal, the signal level during the horizontal and vertical blanking intervals, typically representing zero output.

border region — in a video display, an area of the screen surrounding the *active pixel region*. Many systems allow separately specifying a color value for the border region.

bpp — see *bits per pixel*.

CAV — see *constant angular velocity*.

CD — see *compact disc*.

CD-I — see *Compact Disc-Interactive.*

CD-ROM — see *Compact Disc-Read-Only Memory.*

chroma — see *chrominance.*

chrominance — in an image reproduction system, the signals which represent the color components of the image, such as *hue* and *saturation*. A black-and-white image will have chrominance values of zero. In the NTSC television system, the I and Q signals carry the chrominance information (Sometimes abbreviated as *chroma*).

CLUT — see *color look-up table.*

CLV — see *constant linear velocity.*

coding — the process of representing a varying function as a series of digital numbers.

color balance — in a color video system, the process of matching the amplitudes of the red, green, and blue signals so that the mixture of all three makes an accurate white color.

color bars — a test pattern composed of eight rectangles of different colors: white, yellow, cyan, green, magenta, red, blue, black.

color corrector — equipment for adjustment of the color values of a color video signal. Color correction is usually required for proper reproduction of images from motion picture film.

color differences — the video signals obtained by subtraction of the luminance value from each of the primary color signals.

color look-up table (CLUT) — a table of color values with any bpp format, and indexed by a pixel value of smaller bpp. This allows display of a selected group of colors by a low bpp system, where the group of colors is chosen from a much larger range (the palette) of colors represented by the bpp value used in the *CLUT.*

color mapping — the process of using a *CLUT* for a color display.

color noise — random interference in the color portion of a composite video system. Because of reduced color bandwidth or color subsampling, color noise appears as relatively long streaks of incorrect color in the image.

color subsampling—the technique of using reduced resolution for the color difference components of a video signal compared to the luminance component. Typically the color difference resolution is reduced by a factor of 2 or 4.

color value — the three numbers which specify a color. See also *pixel value*.

compact disc (CD) — the 12 cm.(4.75 in.) optical read-only disc use for digital audio, data, or video in different systems.

Compact Disc-Interactive (CD-I) — an interactive audio/video/computer system developed by Sony and Philips for the consumer market.

Compact Disc-Read-Only Memory (CD-ROM) — an adaptation of the Compact Disc for use with general digital data.

companding — a process of nonlinear *quantization* used to improve signal-to-noise ratio when digitizing with a limited number of bits per sample.

component video — three color video signals that describe a color image. Typical component systems are R,G,B; Y,I,Q; or Y,U,V.

composite video — a color video signal that contains all of the color information in one signal. Typical composite television standard signals are NTSC, PAL, and SECAM.

compression — a digital process that allows data to be stored or transmitted using less than the normal number of bits. *Video* compression refers to techniques that reduce the number of bits required to store or transmit images.

constant angular velocity (CAV) — in optical disc storage systems, refers to drives that maintain the disc rotational speed constant. This facilitates random access, but it reduces storage capacity.

constant linear velocity (CLV)— in optical disc storage systems, CLV refers to drives that adjust disc rotational speed so that the track velocity under the reading head remains constant. This provides the maximum data storage capacity, but it makes random access more difficult.

contouring — in a digital system, the appearance of patterns in a digitized image because the *quantization* did not have enough levels.

critical section — in the DVI RTX software, a critical section is a segment of code in one RTX task which must not be interrupted by another RTX task.

cyan—the color obtained by mixing equal intensities of green and blue light. It is also the correct name for the subtractive primary color usually called "blue."

data rate—the speed of a data transfer process, normally expressed in bits per second or bytes per second. For example, the data rate of CD-ROM is 150,000 bytes per second.

decibel (dB)—a logarithmic unit for expression of ratios. It is based on the generic unit **bel**, 1 bel equals a 10:1 power ratio. Decibel is 1/10 bel. Thus 10 dB also is a 10:1 power ratio, 20 dB is 100:1 power ratio. Since voltage ratios go as the square root of power ratios, it takes 20 dB for a 10:1 voltage ratio or 40 dB for 100:1 power.

device independent — in computer graphics software, an interface specification that does not depend on the characteristics of any particular graphics display device. Also used to refer to a similar hardware independence in any software, not just graphics.

differential PCM (DPCM)—a digital system where the data transmitted or stored represents the difference between data elements (for images: pixels), rather than the data elements themselves.

digital video—video where all the information representing images is in some kind of computer data form, which can be manipulated and displayed by a computer.

digitizing — the process of converting an analog signal into a digital signal. With images, it refers to the processes of *scanning* and *analog to digital conversion*.

discrete cosine transform — a video compression technique.

display list — for 82750 PA, a list of commands placed in VRAM by the host CPU. 82750 PA interprets and executes the display list commands independently from the activity of the host CPU. This is how the two processors communicate while running in parallel.

DRAM — dynamic RAM, read/write random-access memory. This is the standard type of RAM used in personal computer main memory. Dynamic RAM requires a periodic refresh operation in order to retain its data.

driver — a software entity that provides a software interface to a specific piece of hardware. For example, the DVI video driver provides software access to the video board hardware.

dual-ported — with respect to video RAMs, having two channels for data input/output. VRAM (specifically) has a parallel port for access by the CPU and a serial port used for output for display monitor refresh.

dub — a copy of a video or audio tape recording made by replaying the original tape and simultaneously recording the signal onto a new tape, which becomes the dub. Also used as a verb meaning to copy a tape.

edge quantization — a digital artifact caused (for example) by an ADPCM algorithm, where contouring errors are greater in the vicinity of high-contrast edges in the image.

edited master — in post production for audio or video, the first copy produced by the process of editing together the original material.

82750 DA — see *i750*.

82750 PA — see *i750*.

environment — with respect to personal computers, everything surrounding the PC, including peripherals and software.

event flags — in DVI technology, the data elements used by RTX to indicate that something has happened.

file standard — a standard for access of file-oriented data objects on a storage medium. For CD-ROM, the file standard is specified by ISO 9660.

flicker — visible fluctuation of brightness of an image, often a problem in CRT displays if the vertical scan rate is lower than about 50 Hz.

Fourier transform — a process which determines the frequency domain parameters of a series of data points.

frame — the result of a complete scanning of one image. In motion video, the image is scanned repeatedly, making a series of frames.

frequency interleaving — in the NTSC or PAL television systems, the technique of choosing the color subcarrier frequency so that the chrominance frequency components of the signal fall between the luminance frequency components of the signal.

fringing — in a color display, the effect caused by incorrect superimposition of the red, green, and blue images. Incorrect colors appear at the edges of objects in the image. Also called color fringing.

full-motion video — video reproduction at 30 frames per second for NTSC-original signals or 25 frames per second for PAL-original signals.

gamma — in video systems, refers to the gray scale reproduction or amplitude characteristic of the system.

gamma correction — the process of correcting the amplitude transfer characteristic of a video system so that equal steps of brightness or intensity at the input are reproduced with equal steps of intensity at the display.

generation — in electronic recording, one record and replay process. In video production and postproduction, many generations may be required, causing concern for accumulated distortions because of the repeated recording and replay steps.

genlocking—the process of synchronization to another video signal. It is required in computer capture of video to synchronize the *digitizing* process with the scanning parameters of the video signal.

Green Book — the formal standards document for *CD-I*.

Hadamard transform—a video compression technique.

handle variable — a variable that identifies a data object. Often a handle is an index into a table of pointers, where each pointer points to a different data object or to the specific parameters needed to access a data object in mass storage.

hard-wiring — the process of building an algorithm or a technique specifically into hardware.

header file — in C programming, a file that contains definitions and declarations for some specific purpose. Header files traditionally have a .h file name extension.

High Sierra Group—the group that developed the first version of a file standard for CD-ROM. Later the High Sierra standard formed the basis for the ISO 9660 standard.

high-peaker—in a television camera, a circuit that equalizes the high-frequency response of a pickup device or sensor.

hook routine — in DVI runtime software, a routine intended to be run at specific times during the execution of AVSS.

horizontal blanking interval — the period of time when a scanning process is moving from the end of one horizontal line to the start of the next line.

horizontal market — a market (group of customers) that spans many different end use areas on the basis of some kind of common need.

horizontal resolution — the specification of resolution in the horizontal direction, meaning the ability of the system to reproduce closely spaced vertical lines.

hue — the "color" of a colored point, as red, green, yellow, violet.

hue-saturation-intensity (HSI) — a *tri-stimulus* color system based on the parameters of *hue*, *saturation*, and intensity (*luminance*). Sometimes called HSV, for hue-saturation-value.

Huffman coding — a *bit assignment* technique which assigns data values having the highest probability of occurrence to the shortest bit stream words. Less probable data values are given longer bit stream words.

I — in NTSC video, I is one of the two color difference signals, I and Q. The letter I stands for *in-phase*.

IEC — acronym for "International Electrotechnical Commission" — an international standardization body.

image — a still picture, or one *frame* of a motion sequence.

image file — a file of data which represents an image.

image plane — in digital video display hardware which has more than one video memory array contributing to the displayed image in real time, each memory array is called an image plane. See also *bit plane*.

image processing — techniques which manipulate the pixel values of an image for some particular purpose. Examples are: brightness or contrast correction, color correction, changing size (scaling), or changing shape of the image (warping).

in-phase — in NTSC video, refers to being at 0 degrees with respect to the color subcarrier.

interactivity — the ability of a user (or a computer) to control the presentation by a multimedia system, not only for material selection, but for the way in which material is presented.

interlace — in *scanning*, the technique of using more than one vertical scan to reproduce a complete image. In television, 2:1 interlace is used, giving two vertical scans (fields) per frame; one field scans odd lines, and one field scans the even lines of the frame.

inverse transform — in *transform* coding, the process of restoring the original data from the transformed data.

IRE units — a relative scale for measurement of analog video signal levels where blanking level is 0 IRE units and peak white level is 100 IRE units. (IRE is an acronym for *Institute of Radio Engineers*, one of the forerunners of today's worldwide electrical engineer's professional society, the *Institute of Electrical and Electronics Engineers*.)

i750 — the two-chip set which implements DVI Technology on a personal computer. Consists of the 82750 PA pixel processor and the 82750 DA display processor.

ISO — acronym for "Industrial Organization for Standardization" — an international standardization body.

JPEG — acronym for "Joint Photographic Expert Group" — a working party of the ISO — IEC Joint Technical Committee 1, working on algorithm standardization for compression of still images.

keying — in a video system, the process of inserting one picture into another picture under spatial control of another signal, called the keying signal.

lands — in optical recording, refers to the areas of the data tracks which are between the *pits*. These are typically the areas not touched by the recording laser beam during mastering.

layering — in music or sound production, the technique of combining many sound generators to create a richer sound.

level — in video, the signal amplitude.

line — in *scanning*, a single pass of the sensor from left to right across the image.

line pitch — see *pitch*.

line-scan pickup device — a type of solid-state video pickup device which electronically scans only in one direction. *Scanning* in the other direction is accomplished mechanically by relative motion between the pickup device and the image.

linear matrix transformation — the process of transforming a group of **n** signals by combining the signals through addition or subtraction. Used, for example, to convert RGB into YUV.

linear play — playback of a recorded sequence from start to finish without interactivity.

linearity — in television, usually refers to the geometric accuracy of *scanning*. However, linearity is also sometimes used to refer to the accuracy of gray scale reproduction (linearity of the amplitude transfer characteristic), but it is less confusing to use the word *gamma* for the gray scale characteristic.

luminance — in an image, refers to the brightness values of all the points in the image. A luminance-only reproduction is a black-and-white representation of the image.

magenta — the color obtained by mixing equal intensities of red and blue light. It is also the correct name for the subtractive primary color usually called "red."

mastering — in optical recording, the original optical recording process.

matrix transformation — in analog color video, the process of converting the color signals from one tri-stimulus format to another, as for example RGB to YUV.

microcode — a method of programming a microcomputer chip where each program word controls parallel functions of the chip. Microcode program word length is determined by the number of individual functions of the chip to be controlled and is not related to data word lengths.

MIDI — see *Musical Instrument Digital Interface*.

MIPS — Million Instructions Per Second.

motion compensation — a video compression technique that makes use of the redundancy between adjacent frames of motion video.

motion video — video which displays real motion. It is accomplished by displaying a sequence of images (*frames*) rapidly enough that the eye sees the image as a continuously moving picture.

MPEG — acronym for "Motion Picture coding Expert Group" — a working party of the ISO — IEC Joint Technical Committee 1, working on algorithm standardization for compression of motion video.

multitasking — in a computer, a technique that allows several processes to appear to run simultaneously even though the computer has only one CPU. Multitasking is done by sequentially switching the CPU between the tasks, usually many times per second.

multiburst — a test pattern for testing horizontal resolution of a video system. It consists of sets of vertical lines with closer and closer spacing (an example is shown in Figure 2.6.)

Musical Instrument Digital Interface (MIDI) — a serial digital bus standard for interfacing of digital musical instruments. MIDI is widely used in the music industry.

NTSC — acronym for "National Television Systems Committee," the standardizing body which in 1953 created the color television standards for the United States. This system is called the NTSC color television system.

Nyquist limit — in *sampling*, the highest frequency of input signal that can be correctly sampled. The Nyquist limit is equal to half of the sampling frequency.

optical — in the motion picture industry, the name for *special effects*.

overscanned — in a television receiver, the practice of scanning a little beyond the edges of the screen so that the edges of the raster are not visible.

PAL — acronym for "Phase Alternation Line," which is the key feature of the color television system developed in West Germany and used by many other countries in Europe. This system is called the PAL system.

palette — when used with color look-up tables, the range of colors from which the table colors can be selected. The number of colors in the palette is equal to 2 raised to the (number of bits in a CLUT entry) power.

path table — a table in RAM containing CD-ROM directory information.

PC (personal computer) — a software-controlled system containing general-purpose computer functions and intended for use by one user.

PCM — see *pulse code modulation.*

pit — in optical recording, a microscopic depression made on the disc surface by the recording laser beam. Recorded information is contained in the pits and the spaces between pits — *lands.*

pixel — a single point of an image, having a single *pixel value.*

pixel operation — the process of modifying a *pixel value* for some specific purpose.

pixel value — a number or series of numbers that represent the color and luminance of a single *pixel.* Also *color value.*

pixellation — in a digital image, a subjective impairment where the pixels are large enough to become individually visible.

plane — see *image plane* or *bit plane.*

platform — in computers, the base architecture of a computer system. Typical computer platforms are PC AT or Macintosh.

PLV — see *Presentation-Level Video.*

postproduction — in video (and audio), the process of merging original video and audio from tape or film into a finished program. Postproduction includes editing, special effects, dubbing, titling, and many other video and audio techniques.

Presentation-Level Video (PLV) — in DVI Technology, the highest-quality video compression process, PLV requires the use of a large computer for compression.

primary color — in a *tri-stimulus* color video system, one of the three colors mixed to produce an image. In additive color systems, the primary colors are red, green, and blue. In *subtractive* color systems, the primaries are cyan, magenta, and yellow.

primitives — in computer graphics, a set of simple functions for drawing on the screen. Typical primitives are rectangle, line, ellipse, polygon, etc.

production — in video, refers to the process of creating programs. In more specific usage, production is the process of getting original video onto tape or film, ready for *postproduction.*

pulse code modulation (PCM) — in this book, refers to the digitization technique that uses fixed frequency sampling and linear quantization.

pulse cross — a special television monitor mode for testing; the display is synchronized so that the synchronizing pulse portion of the video signal shows in the center of the display.

Q — in NTSC video, Q refers to the *quadrature* color difference signal. It is 90 degrees out of phase with the color subcarrier.

quadrature — being at 90 degrees to the reference.

quantizing — the process of converting an analog value into a digital value having a limited number of bits. This results in reduction of the continuous-level scale of an analog signal to a discrete number of quantizing levels represented by 2^n where n is the number of bits.

quantizing noise — the digital *artifact* caused by *quantizing* with too few levels.

RAM dump — the process of outputting the contents of a contiguous block of RAM as a sequence of bytes.

random-access — in digital memory or mass storage, the ability to access to any point or address without any limitation.

raster — the pattern of motion used in scanning, usually left to right, and repeated over the image from top to bottom as a series of horizontal lines.

raster operation — the process of performing a logical operation on a pixel value. Also called **raster op**.

real images — images captured from nature, usually by photography, cinematography, or television camera; also **realistic** images.

Real-Time Video (RTV) — in DVI technology, the video compression/ decompression technique which operates in real time using the DVI system itself. It provides picture quality suitable for application development purposes, but it will normally be replaced by Presentation Level Video for the final application.

Red Book — the formal standards document for CD Digital Audio.

registration — in video cameras or video displays, the process of causing the three color images to exactly coincide in space.

resolution — the ability of an image reproducing system to reproduce fine detail. In television, resolution is specified in *lines per picture height*, which is the total count of black and white lines that can be reproduced in a distance equal to the picture height.

resolution wedge — a test pattern for resolution of a video camera system consisting of a group of lines angled so that they come closer together as you move across the pattern. An example of resolution wedges for both horizontal and vertical resolution is shown in Figure 2.7.

retrieval engine — a system embodying software or hardware or both for accessing indexed data from a large mass store such as a CD-ROM.

RGB — acronym for red, green, blue.

rms (root-mean-square) — a statistical averaging process where the average is calculated as the square root of the sum of the squares of the values. In video or audio systems it is used to express the magnitude of random noise signals.

rotational delay — in a disc storage system, the time it takes for the desired data to come under the read head, after the *seek* to the correct track has completed.

RTV — see *Real-Time Video*.

run-length coding — a data compression technique that takes advantage of repeated data elements having the same value. Instead of repeatedly coding the same value, the value is coded once along with a count of the number of times to repeat that value.

sample and hold — the technique of *sampling* of a signal and holding that value at the output until the next sample is taken.

sampling — the process of reading the value of a signal at evenly spaced points in time.

sampling rate — the clock frequency for *sampling*, or the number of samples per second.

saturation — the depth of color intensity. Zero saturation is white (no color), and maximum saturation is the deepest or most intense color possible.

scaling — a process for changing the size of an image.

scanner — a device which performs *scanning*.

scanning — the process of converting an image to an electrical signal by moving a sensing point across the image usually in a pattern from left to right, and repeated from top to bottom as a series of horizontal lines.

scanning line — see *line*.

SECAM — acronym for *Sequential Coleur Avec Memoire* (sequential color with memory), which is the color television system developed in France and used in certain other countries.

sector — in a mass storage device, the smallest physical unit of storage.

seek time — in a mass storage device, the time required to position the read head over the track containing the desired data.

sensor (transducer) — a device for converting sounds or images to electrical signals: microphone or video camera.

signal-to-noise ratio (S/N) — in analog video systems, the ratio between the peak-to-peak black-to-white signal and the rms value of any superimposed noise. In analog audio systems, S/N refers to the ratio of rms signal to rms noise.

silicon compilation — a VLSI chip design technique that takes logic diagram input and delivers a complete chip layout and detail design, ready for mask-making.

slope overload — in a data compression/decompression system, the situation where the input signal changes too rapidly to be correctly reproduced by the system. A typical *artifact* of *DPCM* systems.

smear — an analog *artifact* where vertical edges in the picture display a spreading to the left or right. Typically caused by midfrequency distortions in an analog system.

SMPTE time code — a standard for a signal recorded on video tape to uniquely identify each frame of the video signal. It is used for control of editing operations. (SMPTE stands for *Society of Motion Picture and Television Engineers.*)

special effect — a video manipulation technique used to enhance or smooth a transition between camera shots or to create an unusual appearance. Typical transition effects are wipes, fades, or dissolves.

stamper — in optical replication, the stamper is a negative copy of the original master, used directly to press copies for distribution.

statistical coding — in video compression, a coding technique that makes use of the fact that all pixel values are not equally probable.

step size — in a *DPCM* or *ADPCM* system, the amount of output amplitude change represented by the least-significant bit of the DPCM signal. In an ADPCM system, the step size is typically changed dynamically in response to the needs of the signal being processed.

streaking — an analog *artifact* where bright objects in the picture cause a shifting of level that extends horizontally all the way across the picture. Typically caused by distortions to the frequency components below the horizontal line frequency.

stream — the flow of data as a sequence of bits. Also *bit stream*.

stroke font — see *vector font*.

structured graphics — in computer graphics, the process of drawing by means of a limited set of simple functions or *primitives*.

subcarrier — in a composite color television system, a high-frequency carrier on which the chrominance information is modulated, before combining with the luminance signal.

subtractive color system — color reproduction by mixing appropriate amounts of color paints on white paper. The color paint primaries are "red," "blue," and yellow. Note that "red" as used in painting is technically a *magenta* color, and "blue" is technically a *cyan* color.

surrogate travel — the ability to explore a remote or inaccessible site by interactive viewing of material from that site, usually in the form of frame sequences or motion video.

sync information — in television, refers to the part of the video signal devoted to making sure that the display scanning will move in synchronism with the camera scanning.

synchronization time — in a mass storage device, the time required to recognize the data pattern of a track, starting from the time the read head reaches the correct track.

synthetic video — video images created by a computer. A preferred technique is to render images from a computer model by filling the polygons of the computer model with video textures or patterns.

system bus — in a computer system, the parallel data and address path connecting the CPU, memory, mass storage, and I/O devices.

TBC — see *time base corrector*.

telecine — refers to equipment used for television reproduction of motion picture film or film slides.

terminate and stay resident (TSR) — a class of software program which, upon execution, installs itself in RAM and quits. The "resident" module which remains in RAM performs some kind of function for other programs which are subsequently run. The TSR is one approach for installing a special-purpose driver.

threshold — in a digital circuit, a dividing line between circuit signal levels representing different digital values.

time base corrector — equipment that corrects for *time base errors* in video tape recorders.

time base errors — in video tape recorders, analog artifacts caused by nonuniform motion of the tape or the tape head drum. Typically visible as a horizontal jitter or instability in the reproduced picture.

tint — see *hue*.

transducer (sensor) — a device for converting sounds or images to electrical signals: microphone or video camera.

transform — in data compression, a process that converts a block of data into some alternate form that is more convenient or efficient for a particular purpose.

transition effects — *special effects* which occur at the transition between different video camera shots.

transparent — an electronic system is said to be transparent if its output is indistinguishable from its input.

treatment — a narrative writeup, usually nontechnical, which describes a proposed creative work such as a software application or an audio/video production segment.

tri-stimulus — in color reproduction, the method that uses three primary colors or three color signals for image transmission and reproduction.

truncation — in video compression, the technique of reducing the number of bits per pixel by throwing away some of the least significant bits from each pixel.

TSR — see *terminate and stay resident.*

U,V — the scaled-in-amplitude chrominance components of the PAL color television system.

underlay data — in DVI technology, data contained in a separate stream in an AVSS file, intended for use for any purpose which requires data retrieval in synchronism with the frames of video.

underscanned — scanning of a display so that the edges of the raster are visible.

VAPI — acronym for "Video Application Programming Interface" — the C-language programmer's interface for DVI Technology.

VDP — see *video display processor.*

vector font — a font for text characters where the character description is in the form of lines or equations which define the character outline. Also called stroke font or outline font. Vector fonts have the advantage that they can be scaled to any size at the time of display. Their disadvantage is that they take a lot of processing power at the time of display.

Appendix D

Bibliography

1. D. F. Dixon, S. J. Golin, and I. H. Hashfield, "DVI Video Graphics," *Computer Graphics World*, July 1987.

2. D. F. Dixon, M. J. Keith, "Warping Video to 3D Graphics," *Computer Graphics World*, September 1987.

3. R. N. Hurst and A. C. Luther, "DVI: digital video from a CD-ROM," *Information Display*, April 1988, pp. 8–10.

4. A.C. Luther, "You are there... and in control," *IEEE Spectrum*, September 1988, pp. 45-50.

5. A. N. Netravali and J. O. Limb, "Picture Coding: A Review," *Proceedings of the IEEE*, Vol. 68, No. 3, March 1980, pp. 366-406.

6. Philips International, *Compact Disc-Interactive, A Designer's Overview*. New York: McGraw-Hill, 1988.

7. Steve Lambert and Suzanne Ropiequet (ed.), *CD-ROM The New Papyrus*. Redmond,WA: Microsoft Press, 1986.

8.   Suzanne Ropiequet (ed.), *CD-ROM Optical Publishing*. Redmond, WA: Microsoft Press, 1987.

9.   K. Blair Benson (ed.), *Television Engineering Handbook*. New York: McGraw-Hill, 1986.

10.   J. D. Foley and A. Van Dam, *Fundamentals of Interactive Computer Graphics*. Addison-Wesley, 1982.

11.   Intel Corporation, *iAPX 286 Programmer's Reference Manual*. Santa Clara, CA: Intel Corporation, 1985.

12.   N. S. Jayant and P. Noll, *Digital Coding of Waveforms*. Englewood Cliffs, NJ: Prentice-Hall, 1984.

13.   B. W. Kernighan and D. M. Ritchie, *The C Programming Language*, second edition. , Englewood Cliffs, NJ: Prentice-Hall, 1988.

14.   Jack Purdum, *C Programming Guide* Que Corporation. Indianapolis, IN: Que Corporation, 1983.

15.   Mitchell Waite, Stephen Prata, and Donald Martin, *C Primer Plus*, Indianapolis, IN: Howard W. Sams, 1987.

16.   K. Blair Benson (ed.), *Audio Engineering Handbook*. New York: McGraw-Hill, 1988.

17.   C. Sherman (ed.), *The CD-ROM Handbook*. New York: McGraw-Hill, 1988.

Index

Software Diskette Offer

The software examples in this book are available on diskette. A two-disk package containing all source code, make files, and sample images is available from the address below for $24.95. Executable programs are also included, but they are specific to the current version of the DVI Libraries and may not run with future versions. To be sure of a proper match with the DVI version you are using, you should re-compile and link from the source code.

Digital Video Software Offer
Post Office Box 6069
Alexandria, VA 22306-6069

Please specify whether you want 3 1/2" or 5 1/4" diskettes and include your mailing address. Allow 3 - 4 weeks for delivery.

Checks or money orders in U.S. dollars should be made payable to "Digital Video Software Offer". Credit cards cannot be accepted.